D1256071

The Outsourcing Process

Outsourcing has become an increasingly important issue for many organisations. This book provides a framework for an up-to-date understanding of the outsourcing process and the key issues associated with it. It integrates a number of contemporary topics including benchmarking, buyer–supplier relationships, organisational behaviour, competitor analysis, and technology influences. The analysis draws upon both empirical research and real case studies. The author starts by providing guidelines as to when outsourcing is appropriate and what its implications will be, before moving on to explain how outsourcing is implemented. The benefits of both successful outsourcing and the risks and consequences of outsourcing failure are outlined. The book is ideal for use by postgraduate students studying the area of outsourcing. It would also benefit industry managers who are considering outsourcing or who already have outsourcing programmes in place.

Ronan McIvor is a Reader at the University of Ulster.

The Outsourcing Process

Strategies for Evaluation and Management

Ronan McIvor

University of Ulster

CAMBRIDGE
UNIVERSITY PRESS

658. 4058
M 15 r

CAMBRIDGE UNIVERSITY PRESS
Cambridge, New York, Melbourne, Madrid, Cape Town, Singapore, São Paulo

CAMBRIDGE UNIVERSITY PRESS
The Edinburgh Building, Cambridge, CB2 2RU, UK

Published in the United States of America by Cambridge University Press, New York

www.cambridge.org
Information on this title: www.cambridge.org/9780521844116

© R. McIvor 2005

This book is in copyright. Subject to statutory exception
and to the provisions of relevant collective licensing agreements,
no reproduction of any part may take place without
the written permission of Cambridge University Press.

First published 2005

Printed in the United Kingdom at the University Press, Cambridge

A catalogue record for this book is available from the British Library

ISBN-13 978-0-521-84411-6 hardback
ISBN-10 0-521-84411-8 hardback

Cambridge University Press has no responsibility for
the persistence or accuracy of URLs for external or
third-party internet websites referred to in this book,
and does not guarantee that any content on such
websites is, or will remain, accurate or appropriate.

Contents

University Libraries
Carnegie Mellon University
Pittsburgh PA 15213-3890

Figures

Tables

Illustrations

Acknowledgements

I would like to thank the many people who have helped in the writing of this book. The rationale and structure of the book was greatly influenced by empirical research and numerous discussions with academics and practitioners. In particular, a number of colleagues including Paul Humphreys, Eddie McAleer, Tony Wall and Marie McHugh provided valuable feedback on the structure and early versions of certain chapters of the book. Many of the strategies and concepts in this book are based on a number of literature streams including business strategy, economics, organisational behaviour, and inter-organisational relationships. I would like to thank those academics and researchers whose work I have cited throughout the book. Students at the University of Ulster provided valuable feedback on various ideas and concepts in the book and also contributed to the development of a number of the case illustrations. I am also indebted to the companies that allowed me to carry out in-depth case studies to validate many of ideas and concepts introduced. In particular, I would like to thank the people in the organisations that assisted in the development of the case study presented in Chapter 11. The four anonymous reviewers of an earlier draft of the book provided valuable guidance and suggestions for improvement. Eric Willner and Emily Yossarian at Cambridge provided helpful advice and useful comments at various stages in the writing process. Finally, I would like to thank my family and in particular, Deirdre who contributed support and encouragement throughout the preparation of this book.

1   Introduction

A key issue to have emerged for many organisations has been the growing importance of outsourcing. The drive for greater efficiencies and cost reductions has forced many organisations to increasingly specialise in a limited number of key areas. This has led organisations to outsource activities traditionally carried out in-house. Outsourcing involves re-drawing the boundaries between the organisation and its supply base. Although, the term outsourcing has become in vogue in the last few years, organisations have always made decisions on determining the boundary of the organisation. However, the increasing prevalence of outsourcing has led to the concept receiving a significant amount of attention from both academia and practitioners. Outsourcing has progressed from involving only peripheral business activities towards encompassing more critical business activities that contribute to competitive advantage. On the one hand, outsourcing can involve the transfer of business support functions such as cleaning or security to external suppliers in order to obtain higher levels of performance at a lower cost with relatively little upheaval for the organisation. On the other, it can lead to major organisational change that involves dismantling the traditional structure of the organisation, transferring staff to external suppliers, re-defining staff terms and conditions and altering the expectations of the employees that remain within the outsourcing organisation. As a result, outsourcing has become an increasingly important and complex issue for many organisations.

Outsourcing often provokes contrasting reactions from a range of organisational stakeholders including business leaders, unions, employees, politicians and governments. For example, many business leaders regard outsourcing as a powerful vehicle to achieve performance improvements, whilst unions regard outsourcing as another weapon in the armoury of powerful businesses to further erode the terms and conditions of an already embattled employee. Governments and politicians have also become involved in both the debate and practice of outsourcing. Outsourcing has become an extremely political issue in developed economies such as the US and the UK as organisations increasingly outsource many production- and service-related activities to developing economies in order to avail of lower labour rates and more favourable employment legislation. Many governments in

developed economies have also been employing outsourcing as a means of reducing the scale of large public sector organisations and accessing the capabilities of
product and service providers in the private sector.

Where outsourcing is evaluated and managed appropriately it can be a very
powerful means for achieving performance improvements and contributing to the
strategic development of an organisation. Organisations can benefit greatly from
accessing the capabilities of suppliers in a range of areas including catering,
security, design, manufacture, marketing, logistics and information technology.
In fact, the development of supply markets for a range of business services has
precipitated the trend towards outsourcing and provided opportunities for organisations to enhance their own competitive position. However, many organisations
have failed to achieve the desired benefits associated with outsourcing and
experienced the consequences of outsourcing failure. Such a trend has increased
the likelihood of outsourcing suffering a similar backlash to those that have
followed Total Quality Management and Business Process Re-engineering.
Similarly, concepts such as outsourcing tend to be adopted by practitioners
under pressure to demonstrate knowledge and expertise but who do not have
the time to assess whether the concept is applicable to their particular business
situation. Indeed, many outsourcing failure cases have been as a result of the
misapplication of the concept by practitioners. Organisations that have experienced poor results with outsourcing often have had limited knowledge and experience of outsourcing. Organisations have often embarked upon extensive
outsourcing without fully understanding the concept. For example, organisations
have outsourced activities with which they are having problems without diagnosing
the causes of poor internal performance. Rather than attempt to effect performance
improvements internally, they have rushed into outsourcing with scant consideration of the consequences. Moreover, opportunistic suppliers can be extremely
adept at holding organisations to ransom through locking them into long-term
contracts and adding supplementary charges as conditions change. Suppliers often
exploit the naiveté on the part of the outsourcing organisation to negotiate a
sufficiently robust contract.

Initial research for the writing of this book revealed that organisations were
experiencing many of these problems. Organisations were encountering difficulties in both the evaluation of the suitability of outsourcing for their organisation
and in the management of the outsourcing process. These difficulties stemmed
from the lack of a structured approach to the evaluation and management of
outsourcing. Furthermore, through extensive research and teaching in this area, I
have found that there is no single text that adequately addresses the topic of
outsourcing. Many books or individual articles are generally oriented towards a
single perspective and do not provide an overall integrative framework necessary
to understand outsourcing evaluation and management. The objective of this

book is to summarise and integrate the latest research in outsourcing and related disciplines in a way that is accessible to both students and practitioners and in a way that facilitates its application. The book synthesises much of the literature that relates to outsourcing into a comprehensive framework for understanding the process of outsourcing evaluation and management. In particular, the outsourcing framework is influenced by literature from a number of areas including business strategy, economics and inter-organisational relationships. The implications of a number of key influences on the outsourcing process such as organisational capability, supplier capability, competitor actions and supply market risk are examined and a number of alternatives are offered based upon an analysis of these key influences. The book is influenced by empirical research that involved talking directly to practitioners in order to elicit their views on key aspects of the outsourcing process. In-depth case studies were also carried out with a number of companies that had embarked upon extensive outsourcing. The framework has been tested and refined by many presentations and discussions with both academics and practitioners. Much of the structure of the book is influenced by the stages involved in outsourcing evaluation and management. The book is structured as follows:

- *Chapter 2: The trend towards outsourcing* – this chapter provides an overview of the outsourcing concept. It identifies a number of developments in the business environment that have led to the increasing prominence of outsourcing. The potential benefits and risks associated with outsourcing are outlined along with the inter-organisational relationships that have been adopted as a result of the trend towards outsourcing.
- *Chapter 3: Theoretical influences on outsourcing* – the characteristics and limitations of a number of theoretical influences on outsourcing are outlined. It is argued that there are a number of aspects to each theoretical perspective that can assist in outsourcing evaluation and management.
- *Chapter 4: The outsourcing process: a framework for evaluation and management* – this chapter provides an overview of the practical problems with outsourcing and an overview of the stages involved in the framework for outsourcing evaluation and management. The stages in the framework serve as a structure for the following six chapters in the book.
- *Chapter 5: Determining the current boundary of the organisation* – in this chapter a number of models are presented that can assist in providing an outline of the current scope of an organisation's activities. Outsourcing evaluation and management analysis is carried out at the activity level throughout the book.
- *Chapter 6: Activity importance analysis* – a key element in outsourcing evaluation is assessing the importance level of organisational activities. An analysis of the competitive environment and customer needs can assist in identifying which activities are critical for success in the business environment. A number of

techniques are outlined that can assist in determining the importance level of organisational activities.

- *Chapter 7: Capability analysis* – this chapter provides a structure for assessing the capability of an organisation in activities relative to competitors and potential suppliers for outsourcing evaluation. The relative performance of the sourcing organisation in an activity will have a major influence on whether an activity should be performed internally or outsourced.
- *Chapter 8: An analysis of the strategic sourcing options* – this chapter outlines the implications of each strategic sourcing option based upon the importance level and capability of the organisation in the relevant activity. Issues considered include the performance disparity in the activity relative to competitors or suppliers, technology influences, behavioural considerations and supply market risk.
- *Chapter 9: Developing the relationship strategy* – this stage in the analysis is concerned with outlining the steps involved in developing a relationship strategy that enables the achievement of the outsourcing objectives. A number of potential relationship strategies are discussed that are influenced by both the importance of the activity and the level of supply market risk.
- *Chapter 10: Establish, manage and evaluate the relationship* – in this chapter the stages involved in implementing the outsourcing strategy are outlined. A number of issues are considered including supplier selection, contracting, supplier and relationship evaluation.
- *Chapter 11: Outsourcing experiences at Telco* – this chapter presents a case study documenting the outsourcing experiences of a company in the telecommunications industry. The analysis presented is related to the framework for outsourcing evaluation and management presented in the book.
- *Chapter 12* – this chapter provides a number of practical lessons for outsourcing evaluation and management and also identifies a number of potential future developments.

This book is intended for readers in the academic market, who require an up-to-date understanding of the outsourcing process and a structured framework to evaluate and manage the key issues associated with outsourcing. The book is of interest to students on postgraduate (MBA, MA and MSc) programmes studying the subject of outsourcing. The book can be used as a core text on a stand-alone outsourcing module or on a module such as business strategy, supply chain management, information systems, and operations management where significant attention is given to outsourcing. The book is also of value to students who are researching the area of outsourcing. Chapter 3 provides a detailed and critical evaluation of the key theoretical influences on outsourcing. It is shown that there are a number of inter-dependencies with each of the theoretical influences that can assist in understanding the outsourcing process. In particular, the importance of the business strategy and inter-organisational relationship literature to an understanding of outsourcing is emphasised.

In contrast to many other books on outsourcing, considerable attention is given to evaluating whether outsourcing is appropriate for organisational activities. This is reflected in the title, which emphasises both the evaluation and management aspects of the outsourcing process. Referring to the chapter structure of the book it can be seen that particular emphasis is given to understanding the implications of outsourcing before implementation. Furthermore, in contrast to many other books on outsourcing that focus solely on information technology and outsourcing, this book is generic and applicable to evaluating outsourcing for a range of organisation activities. I have attempted to make the book as interesting as possible to both students and researchers in the area through the use of illustrations and empirical research. The illustrations in each chapter are intended to enrich the analysis and provide support for the strategies proposed. The book is strengthened through reference to contemporary research in the area of outsourcing in a range of leading international journals including the *Strategic Management Journal, Academy of Management Executive, Journal of Management, Administrative Science Quarterly, Long Range Planning, Sloan Management Review, Harvard Business Review*, etc.

Although much of the book is oriented towards the academic community, it is also of value to practitioners who are involved in or considering outsourcing. The book proposes new ways of looking at the outsourcing process, and employing outsourcing as a vehicle to obtain performance improvements and achieve competitive advantage. A framework is provided which employs a number of tools and techniques that many practitioners are already familiar with applying, including competitor analysis, benchmarking, cost analysis and buyer-supplier relationship management. Although the analysis presented employs a number of the theoretical models, these can make a valuable contribution to strategic decision-making. The approach in the book builds in substantial parts of contemporary research and theory. Research and theory can assist in stimulating a deeper understanding of the key issues associated with outsourcing. The book provides guidelines on deciding whether outsourcing is appropriate and if so, how the outsourcing process should be managed. Chapter 12 provides some lessons on the outsourcing process that have been drawn from the analysis presented in the book. It alerts practitioners to the key issues that should be addressed if they are approaching the problem themselves. It will also be useful to companies that are considering employing consultants to undertake both the evaluation and management of outsourcing.

2 The trend towards outsourcing

2.1 Introduction

The increasing prevalence of outsourcing has led to it being considered central to the strategic development of many organisations. Outsourcing is increasingly being employed to achieve performance improvements across the entire business. For example, one particular growth area has been the externalisation of Information Technology (IT) with a recent report showing companies outsourcing 38% of their IT functions to external providers (Barthelemy, 2001). Another area of the business that is also increasingly being outsourced is the human resource function. An American Management Association survey found that 77% of firms surveyed in the US had outsourced a number of human resource activities including training and development, executive recruitment and payroll management (Anonymous, 1997). The outsourcing decision can often be a major influence on the profitability and competitive position of the organisation. However, many organisations possess a limited understanding of outsourcing and in particular the potential benefits and risks and how they should be managed. This chapter provides an overview of the outsourcing concept. It outlines a number of key developments in the business environment that have occurred including globalisation, advances in information technology, reforms in the public sector and more demanding consumers. These changes have forced organisations to be more flexible and responsive to customer needs. As a result, many hierarchically controlled organisations that have previously performed the majority of business activities internally have been forced to create more network-oriented organisational structures, which involves outsourcing activities to specialist suppliers. A number of the key dimensions of outsourcing are identified including the potential advantages of outsourcing, the potential risks, and inter-organisational relationship configurations that have been adopted as a result of the trend towards outsourcing. These issues will be dealt with in more depth in the context of outsourcing evaluation and management in the following chapters.

2.2   An overview of the outsourcing concept

Outsourcing involves the sourcing of goods and services previously produced internally within the sourcing organisation from external suppliers. The term outsourcing can cover many areas, including the outsourcing of manufacturing as well as services. The term 'outsourcing' is most commonly used in relation to the switching of the supply of product or service activities to external suppliers. Outsourcing can involve the transfer of an entire business function to a supplier. Alternatively, outsourcing may lead to the transfer of some activities associated with the function whilst some are kept in-house. Outsourcing can also involve the transfer of both people and physical assets to the supplier. Outsourcing is not just a straightforward financial or purchasing decision. In many cases, outsourcing is a major strategic decision that has implications for the entire organisation. The evaluation and management of the outsourcing process involves a number of important elements. A starting point in the evaluation process involves analysing whether outsourcing an activity is appropriate for the organisation. This involves considering issues such as the capability of the organisation in the activity relative to competitors, the importance of the activity to competitive advantage, the capability of suppliers to provide the activity, the level of risk in the supply market, potential workforce resistance and the impact upon employee morale. Where the decision to outsource has been made, a number of important issues have to be considered including supplier selection, contract negotiation and the transitioning of assets to the supplier. Significant attention should also be given to managing the relationship with the supplier to ensure that outsourcing meets its intended objectives.

A term often used in the context of outsourcing is vertical integration. Vertical integration refers to the level of ownership of activities either *backward* (for example, component manufacture or inbound logistics) into the supply chain or *forward* (for example, distribution or after-sales service) towards the customer or end user of the product or service. This is illustrated on Figure 2.1.

Vertical integration is similar to the outsourcing concept in that it is concerned with the decision on whether to perform an activity internally or source it from an external supplier. Another term that is often used in a manufacturing context is 'make-or-buy', which has been around for many years. The term 'make-or-buy' is associated with the decision on whether to manufacture a component internally or buy it from an external supplier. In fact, it could be argued that make-or-buy is a more appropriate term as it implies that there should be an evaluation of the suitability of either internal or external supply whereas the term outsourcing implies that the decision to use an external supplier has already been made without any consideration of whether it is appropriate for the organisation. Successful

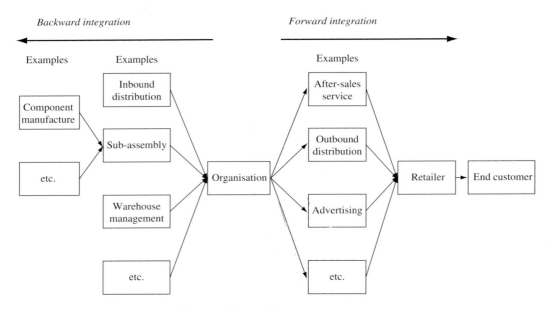

Figure 2.1 Backward and forward integration

application of the outsourcing concept involves an analysis of whether outsourcing is appropriate for the organisation and, if so, how the outsourcing process should be managed.

Organisations have always employed external product and service providers to carry out a range of business activities such as catering, security, distribution, accounting and information technology. However, many organisations are increasingly outsourcing a wider range of activities and a greater level of the value associated with these activities. In effect, organisations are no longer outsourcing peripheral activities alone but extending the scope of outsourcing to encompass more critical activities that contribute to their competitive position. Consequently, the evaluation and management of the outsourcing process has become increasingly complex. For example, contracts are more complex as agreements become more sophisticated in terms of measurement procedures, the financial management of transferred assets and the inclusion of clauses associated with bringing the activity back in-house (Quelin and Duhamel, 2003). Furthermore, the presence of political influences on the decision-making process, such as the vested interests of the functions under scrutiny to keep the relevant activities in-house, can prevent the choice of the most appropriate sourcing option for the organisation. The strategic nature of the decision means that it not only involves input from operational management, but also from top management. Considerable care and effort has to be given to evaluating and managing the outsourcing process. Outsourcing decisions are extremely costly and difficult to reverse. For example, in a study of outsourcing, Barthelemy and Geyer (2000) found that in the event of outsourcing not meeting its intended objectives, it can take an average of 8 to 9 months to switch to another service provider or bring the activity back in-house at the end of the contract.

It is important to distinguish between two different types of outsourcing. Outsourcing can be used to either maintain the competitive position of the organisation or act as a source of competitive advantage. Normally, when organisations approach outsourcing for the first time they are using it to reduce costs and improve performance in a particular activity. Clearly, outsourcing to reduce costs can deliver benefits for the organisation and impact directly upon the bottom line. However, outsourcing employed primarily at this level is only likely to maintain the competitive position of the organisation. In many cases, it is likely that the organisation is accessing the capabilities of a supplier that are also accessible to its competitors. Indeed, many organisations outsource as a result of following the lead of a competitor. Outsourcing for the sake of outsourcing or responding to competitor actions is not a basis for creating competitive advantage. However, organisations can employ outsourcing as a vehicle to achieve competitive advantage. Studies have shown that organisations are becoming more ambitious and sophisticated in both their objectives and approach to the outsourcing process. For example, Saunders *et al.* (1997) found that in many cases the major motives for outsourcing were technological and strategic and not cost reduction. It was found that outsourcing allowed organisations to achieve a number of strategic benefits ranging from the more rapid adoption of new technologies to being more responsive to customer needs by better coping with variations in product demand. PriceWaterhouseCoopers (1999) have found that outsourcing has moved from searching for efficiencies and improvements in a single process or activity, to reconfiguring entire processes in order to obtain greater value across the organisation.

In effect, these organisations are reacting to changes in the external business environment. Increasingly, the ability of organisations to reduce costs is regarded as a pre-requisite for participation in many industries whereas innovation and responsiveness to customer needs are regarded as potential sources of competitive advantage. Outsourcing strategies are becoming more ambitious in order to meet these challenges. A key element of using outsourcing as a source of competitive advantage is the management of suppliers. Accessing the capabilities of suppliers offers organisations the opportunity to improve their own performance. However, a key aspect of unlocking this potential is the way in which the customer organisation manages the relationship with the supplier. Studies have shown that some organisations can obtain higher levels of performance from their suppliers than their competitors. For example, Liker and Wu (2000) found in a comparison of supplier plants with product lines serving both US car makers and Japanese transplants that the actions of the suppliers' customers significantly affected the ability of suppliers to pursue lean manufacturing principles. Suppliers had much leaner operations within their plants and in logistics processes when dealing with Japanese customers. In order to improve supplier performance, the Japanese companies employed a number of mechanisms including more level production schedules, a disciplined system of delivery time periods, assisting suppliers in developing lean capabilities and the development of a transportation system to handle mixed-load and small-lot deliveries. In fact, astute relationship manage-ment in the value chain has become a competitive imperative. As a result, many vertically integrated organisations have been increasingly outsourcing to create a more virtual configuration that relies on a range of business partners to undertake the major parts of the value chain (Kutnick, 1999).

2.3   Changes in the business environment driving outsourcing

The trend towards the increased use of outsourcing has been driven by a number of inter-related factors in the external business environment.

2.3.1   Globalisation

Over the last few years the external business environment has become increasingly global for many industries. Many organisations are now competing on a global basis. Regional agreements such as the North American Free Trade Agreement (NAFTA) between the US, Canada, and Mexico and the development of the European market with a single currency have facilitated the development of trade on a global basis. Increasingly, there has been a shift away from national markets as distinct entities, protected from each other by trade barriers and distance and time barriers, towards a system in which national markets are merging into a single global market (Hill and Jones, 2001). This trend has led organisations to expand the geographical scope of their business operations in terms of the markets they serve and the production locations for the creation of their products and services. These changes have presented organisations with significant opportunities. For example, companies have been in a position to achieve greater economies of scale, share investments in research and development and marketing across their various markets, and access lower cost labour sources for both the manufacture of their products and delivery of their services. Futhermore, many organisations, such as Microsoft and Sony, have been able to establish internationally recognised brands for their product and service offerings. Throughout the developed nations of the world including the US, Europe and Japan there has been a growing prevalence of standardised consumer products in areas such as electronics, computers and cars. However, the trend towards globalisation has presented organisations with many challenges. For example, establishing a global presence can involve managing and co-ordinating a network of manufacturing, distribution, and retail sources. Differences in language, culture, legal requirements and currency movements can further complicate this process. Companies also have to be responsive to local needs and rapidly react to any changes in these markets.

Increasing deregulation has led to the globalisation of external environments in which many organisations compete. For example, since the 1990s, the media and communications markets in the US and Europe have experienced significant liberalisation and deregulation. Increasing competition along with changing organisational structures has transformed the business environment facilitating global competition (Wirtz, 2001). There has also been increasing deregulation in a range of industries including telecommunications, air travel and utilities. The most significant implication of globalisation for organisations has been the intensifying of competition in many industries. The boundaries of many industries span internationally with competitors existing in both home and international markets. Previously, national markets with three or four established companies and a low level of competition have been transformed. Increasing competitive rivalry has reduced the profits of many organisations and forced them to reduce costs, improve customer responsiveness and quality. For example, these impacts have been very pronounced in the European airline industry. Increasing competition from low cost airlines such as Easyjet and Ryanair, very much driven by deregulation, has forced many established European airlines including British Airways, Swiss-Air and Air France to embark upon radical restructuring programmes in order to eliminate inefficiencies and reduce costs. Increasing competition has compelled organisations to adopt strategies such as the more rapid development of new products and services and the restructuring of business processes in order to reduce costs and eliminate inefficiencies.

Globalisation has also had significant political and economic implications. Indeed, globalisation has provoked furious debate and in some cases protests in both the developed and developing countries of the world. The liberalising of international trade has allowed many countries to grow more quickly through exports. For example, export-led growth has been central to the industrial policy of many Asian countries, which in turn has raised living standards in many of these countries. Proponents of free trade base their arguments on the law of comparative advantage first proposed by the English economist David Ricardo (1772–1823). The law of comparative advantage states that countries should specialise in producing and exporting the products or services that they produce at a lower relative cost to competing countries. However, the liberalisation of trade has led to the loss of many jobs in developed economies as companies increasingly outsource manufacturing- and service-related activities to developing economies with lower labour costs – a trend which is sometimes referred to as off-shoring. Outsourcing has become increasingly associated with globalisation. Many organisations have suffered a backlash as a result of outsourcing activities to developing economies. Rather than being seen as raising the living standards of developing economies through outsourcing, these organisations are being accused of exploiting workers that are being paid wage rates at a fraction of the level relative to their counterparts in the developed economies. Moreover, globalisation is seen as a weapon to further strengthen the Western economies whilst keeping the developing economies in a weak position. Globalisation has come to symbolise the global inequalities and the hypocrisy of the developed economies such as the US and the EU (Stiglitz, 2002). The powerful developed economies have argued for and in some cases, forced the opening of markets in developing economies whilst continuing to keep their home markets closed to the products of developing economies. For example, tariffs imposed on the cotton and sugar exports of developing countries by the US and the EU, along with huge subsidies to producers in the developed economies that reduce commodity prices, are considerable barriers to the development of developing nations (Mathiason, 2004).

Illustration 2.1

Offshoring
Offshoring refers to the transfer of organisational activities carried out locally to product and service providers in other countries. Offshoring has been an important practice for organisations for a long time, particularly in the area of manufacturing. More recently, the trend towards offshoring has been particularly pronounced in business services including areas such as telemarketing services and software development. For example, many organisations in the UK, US and Australia have outsourced much of their call-handling activities ranging from after-sales support to direct marketing. Much of this outsourcing has involved the use of both local call-centres and offshore call centres in a range of locations including India, Malaysia and Jamaica. Organisations have pursued this strategy of outsourcing for a number of reasons including cost savings, high labour turnover in local facilities and access to specialist service providers. In particular, many organisations have outsourced telemarketing-related services to off-shore locations in order to access service providers with much lower labour costs. Labour costs in India account for around 30% of total costs for a call centre whilst in the US or UK they account for around 70% of the total costs. The call centres in these countries with much lower labour rates can typically attract a high number of applications from well-qualified and highly literate

graduates. For example, the pay levels in India relative to other professions has encouraged doctors, lawyers and accountants into doing shifts in call centres. Alternatively, in developed countries call centres have considerable difficulties both recruiting and retaining staff. Furthermore, as a result of significant growth in the service sectors of developed economies such as the UK, companies have had to offshore service activities in order to cope with increasing demand.

Many organisations have decided against outsourcing because they believe that foreign service providers cannot provide comparable levels of service with that of local service providers. Indeed, Dell decided to bring back in-house some customer service activities from India as it had received complaints from its business customers about poor service from operators. Some of the potential savings in labour costs have to be weighed against the additional costs and difficulties associated with managing operations in distant locations. Data privacy and the physical security of foreign sites is another concern. Although most software and call centre work is undertaken through dedicated telecommunications facilities rather than over the Internet, there is the potential threat to consumer privacy and the disclosure of consumer information by foreign employees. Organisations have avoided offshoring activities due to adverse publicity and the potential damage to their reputation. In addition, many service activities cannot be outsourced to developing countries because they involve personal one-to-one interaction. For example, many services offered by retailers, restaurants and hotels are provided locally and cannot be outsourced.

In many cases, the trend towards offshoring has sparked considerable controversy in developed countries with, for example, unions in the UK claiming that it will become 'a nation of hairdressers'. Indeed, some unions in the UK have openly campaigned for the boycotting of companies that engage in offshoring. Some argue that the current trend towards outsourcing activities in the service sector to lower wage economies is similar to the situation that has already occurred in other business sectors such as shipbuilding, coal mining and car production. The issue has also become extremely political with legislation being enacted in developed economies to prevent the movement of jobs offshore. For example, several states in the US have prevented offshore companies from undertaking government contract work. The governor of Indiana cancelled a $15 million contract with an Indian company, even though its bid was $8 million less than that of the most competitive US-based competitor. These actions, in turn, have led to resentment in developing countries with justifiable claims of protectionism and Western double standards on the promotion of free trade.

However, there is evidence to show that both developed and developing countries can benefit from offshoring. Consumers in developed countries can benefit from lower prices for products and services. A study undertaken by the management consultancy McKinsey analysed the practice of offshoring and its wider economic impact in the US. Through its findings it has argued that the benefits of off-shoring outweigh the downsides for the US economy. For example, it found that for every $1 dollar lost to an offshore worker, the US economy gains as much as $1.12 from reinvesting cost savings for growth, supplying additional staff in the foreign locations and allowing call centre employees to focus on higher skilled and better paid jobs. Greater competition from offshore product and service providers can act as a stimulus for local providers to achieve productivity improvements and higher levels of innovation. Due to increasing globalisation and the dismantling of protectionist practices, competition from lower wage countries is likely to precipitate the shift away from low-skilled jobs in the service sectors of developed countries. A major challenge for governments in these countries is to pursue policies that enhance job opportunities for lower-skilled workers. Furthermore, although the offshoring of services has been confined primarily to standard software development and telemarketing services, there are legitimate fears in developed economies that the trend towards offshoring may provide a threat to higher value-added work in the long-term. For example, countries such as China and India have been investing in higher education, particularly in the areas of engineering and communications in order to develop their skills base. The lower labour rates and flexibility of a non-unionised and younger labour force in these countries relative to Western economies has and is likely to further enhance the competitiveness of many of these countries.

Sources: Alden, E., Luce, E. and Merchant, K. (2004). Public opinion in Western economies might not be in favour, but cost differentials and labour flexibility mean there is plenty of scope for further migration of jobs. *The Financial Times*, London, 28 January, p. 19.
Khan, S. (2003). Bombay calling, *The Observer*, London, 7 December, p. 19.
Connon, H. (2003). Is it all bad when UK jobs go to India? *The Observer, Business Supplement*, London, 7 December, p. 19.

2.3.2   Developments in information and communication technologies

Advances in telecommunications and information technology have facilitated the trend towards outsourcing. The increasing importance of innovative information and communication technologies (ICTs) for economies and societies has been attracting considerable attention both from academia and practitioners. The evolution of ICTs has been advancing at a more rapid rate than many earlier technologies. Some have argued that the developments in ICTs can amplify 'brain power' in the same way that the technologies of the industrial revolution amplified 'muscle power' (Woodall, 2000). In recent decades ICTs have deeply affected the way business is performed and the way in which organisations compete (Porter, 2001). Most organisations cannot compete effectively without employing information technology. Information technology (IT) can create efficiencies in a number of business areas ranging from design to marketing to finance, and furthermore IT and globalisation are closely linked. IT has greatly facilitated the ability of organisations to both globalise production and access new product markets. In particular, the advance of the Internet, with its vast range of potential services and applications, has led to a transformation of corporate strategy since the middle of the 1990s, as reflected in the increasingly common use of terms such as industry convergence, virtual corporations and electronic commerce (Picot *et al.*, 1997).

Traditional industry boundaries are blurring as increasingly many industries converge or overlap, especially in IT industries (Bettis and Hitt, 1995). For example, Evans and Wurster (2000), using the private banking industry as an illustration, argue that these technology developments will lead to a convergence of industries and the unbundling of traditional value chains. Moreover, Internet retailers such as Amazon compete across a number of traditional industries by offering a range of products that includes CDs, DVDs and books. These trends have had a marked effect on traditional strategic thinking that considered industry structure as the primary determinant of company profitability. The benefits of the Internet – making information available; reducing procurement, marketing and distribution costs; allowing buyers and suppliers to transact business more easily – also make it more difficult for organisations to profit from these benefits. For example, the Internet allows consumers to search more comprehensively at negligible costs for the most competitive price for a range of products or services. The Internet can alter industry structures in ways that reduce profitability, and it has a levelling effect on business practices, reducing the ability of any organisation to obtain a sustainable competitive advantage (Porter, 2001). A major challenge for management in exploiting the Internet is to realise that these fundamental changes are creating a situation where organisations are operating in an increasingly global, technologically interconnected, and information-driven world (Sampler, 1998).

2.3.3 Public sector reforms

The trend towards increased outsourcing has also been influenced by wide-ranging reforms occurring in public sector organisations in many countries. For example, successive governments in the US and UK have pursued radical public sector reforms which have placed the use of competitive market mechanisms at the heart of these reforms. Proponents of this philosophy argue that assets and activities should be transferred from the public sector to the private sector in order to improve performance. They also argue that the public sector should aspire to levels of performance attained in the private sector. In a study of public sector organisations carried out in a number of countries including the US, the UK, France, Germany, Japan and Australia, Domberger (1998) found that outsourcing has become a significant and increasing practice. Much of the force behind this trend has been the prevailing belief that best value is achieved through the use of competitive market solutions for service provision. For example, the impetus for greater application of market forces to the public sector in the US came from the publication of *Reinventing Government*, which emphasised the benefits of competition and customer choice as a means of delivering better and more cost-effective services to citizens (Osborne and Gaebler, 1992). In the UK, during the 1980s and 1990s, successive governments pursued policies which encouraged free market mechanisms in the public sector and discouraged state intervention where possible. Market mechanisms have also been prevalent in developing countries. For example, in Thailand the utility industries such as electricity have been privatised, which has involved the separation of generation from transmission and distribution under a mixed system of public and private ownership (Cook, 1999).

The most prominent concept associated with using the market in the public sector has been the privatisation of publicly owned organisations. However, the introduction of internal markets and the outsourcing of public service services such as refuse collection, building, catering and cleaning have also become very prominent. In the UK, outsourcing – often referred to as contracting-out in a public sector context – began in local government in the 1980s, spreading to central government in the 1990s, following the *Competing for Quality* White Paper. A key initiative was competitive compulsory tendering (CCT) which required the regular re-tendering and the achievement of a stated minimum rate of return on the value of the contract that involved agreeing a formal contractor– client split. Lonsdale and Cox (2000) argue that these reforms have had the following effects.

- It supported the philosophy that third-party external contractors could provide goods and services more efficiently and effectively than internal departments.
- In many areas, outsourcing by public sector organisations contributed to the development of supply markets for goods and services. For example, this was particularly prevalent in the area of IT where many IT service providers developed their business to offer a range of outsourcing services ranging from IT consultancy to applications management and development.

A further development was the Private Finance Initiative (PFI), which led to the development of a range of infrastructure programmes that were designed to deliver better value for money for public services and create market opportunities for the private sector. The constant drive for cost reduction and the efficient use of resources forced many public sector organisations to consider reducing the scale of government departments and public services. This trend towards radically changing the large hierarchical nature of public sector organisations with more responsive customer-oriented network structures is in common with the changes occurring in many commercial organisations. However, these developments in the public sector have been very controversial. For example, the high level of union-isation in the public sector has hindered the freedom with which governments can pursue such reforms.

2.3.4   More demanding consumers

In many business sectors consumers have become more sophisticated and demanding as they have become more knowledgeable on issues such as price, reliability and availability. Consumers are demanding a more customised product and service at a lower price. Consumers are much more mobile in terms of ease of access to alternative sources of supply as a result of increased competition in many markets and the advent of the Internet. With global access through the Internet to more products and services than ever before and with instant communications, typical constraints, such as time and distance are rapidly disappearing. As a result, the loyalty of consumers to product and service providers is diminishing. As well as expectations continuing to rise, these consumers will become increasingly unpredictable. In the past, consumers determined the value of a product or service on the basis of some combination of quality and price. Treacy and Wiersema (1993) argue that customers will increasingly employ an expanded concept of value that encompasses convenience of purchase, after-sales service, uniqueness and reliability. With consumers becoming more sophisticated and powerful they will no longer settle for whatever companies are offering. Organisations are now being forced to be more responsive to customer needs in a range of areas. In some consumer markets, it is now possible for the consumer to describe exactly what they want by actively modifying and adding features to the product virtually. For example, on Nike's web-site – *nike.com* – the customer can design and choose the features to create their 'favourite' running shoe. Car-makers also allow potential customers to create their preferred car on screen and also access immediate prices for any extra features required.

Traditionally, organisation designers have 'designed' products that they think consumers want. In effect, the organisation has decided what the consumer wants. Coupled with this is the fact that the design process can be a much-compromised process to allow for issues such as the standardisation of components to ensure ease of manufacture. To illustrate, much has been written about the way in which Benetton moved the dyeing process for its garments to a later stage in the production process to allow them to anticipate end-customer demand across the colours (Christopher, 1998). However, allowing the end-customer to be actively involved in the design of the product is taking this a stage further. Rather than organisation designers determining the features that the product should have, the end-customer now has an opportunity to be an active participant in this process. Such changes are not only affecting commercial organisations but are also having major implications for public sector organisations. Consumers are increasingly demanding higher levels of service and responsiveness from public sector organisations. With the consumer demanding a more information-enriched and interactive relationship with commercial organisations, public sector organisations are having to respond with innovations in the way in which they interface with citizens in order to achieve greater accessibility and efficiency. Academic institutions are already using the Internet to interact virtually with students. For example, many academic institutions allow students to carry out activities such as submitting applications for courses, enrolment and access to teaching materials and results. In the past, many public sector organisations have been 'protected' from the harsh commercial reality of competition and an ever increasingly demanding citizen. However, in the future, the expectations of consumers will continue to rise with the increasing use of on-line trading which allows business to be conducted at any time of the day.

2.4   Evolving organisation structures

These changes in the business environment are having major implications for the way in which organisations are structured and managed. The level of change and uncertainty in many industries is likely to accelerate, as organisations have to deliver more with fewer resources. In order to compete effectively in this environment, organisations must not only continue to reduce costs, but also focus on innovative activities that can deliver customers value and create competitive advantage. In fact, changes in the business environment are forcing companies to move away from earlier paradigms of structuring and managing organisations such as the mass production paradigm where organisations preyed on a largely naïve consumer. The way in which many organisations meet the needs of their customers is still heavily influenced by the principles associated with mass production. The earliest version of mass production targeted 'vanilla' products at large volume markets with little attempt to further satisfy the needs of a largely ignorant consumer. The organisation could afford to adopt such a strategy because the consumer was less mobile in terms of ease of access to alternative sources of supply. Although the term mass production implies the manufacture of physical products, its principles have been prevalent in other areas such as travel, education and banking. For example, the majority of opening hours – especially in the areas of banking, education and health – remained limited to times of the day that were inconvenient to many potential users of these services. However, gradually the principles of mass production have become redundant in many industries due to factors such as innovations in technology, globalisation, greater levels of competition, and a more sophisticated customer. Organisations have had to become more proactive in achieving innovations that directly benefit the consumer rather than the organisation.

As a result of these changes in the business environment, organisations have begun to radically rethink their organisational structures. Historically, companies have chosen either a strategy of vertical integration (make internally) or outsourcing (source externally) to bring their products to the market, based on the relative benefits and disadvantages of each alternative. During the 1980s and 1990s, many highly vertically integrated companies have divested large parts of their business in order to focus and specialise in a limited number of core areas. Large vertically integrated organisations have been found to be incapable of achieving greater levels of efficiency and of being responsive to an ever more demanding customer in a rapidly changing business environment. Consequently, many of these organisations have been attempting to bridge the gap between the traditional make versus buy decision by combining the strengths of vertical integration and outsourcing in order to achieve vertical integration without financial ownership. This trend has precipitated changes in organisation structures that have been occurring across many business sectors. In a stable business environment with low levels of change, organisations tend to use a mechanistic structure characterised by a bureaucratic and hierarchical chain of command. Conversely, in a dynamic, rapidly changing business environment, organisations tend to adopt a more organic structure characterised by high levels of responsiveness and a flattened hierarchy. Such organisations rely on internal networking through the development of an informal culture and the limiting of functional barriers. As a result, many organisations have been forced to adopt more organic structures in order to achieve greater levels of efficiency and cost reductions. These developments have given rise to the use of a range of terms to describe evolving organisational structures such as the 'network organisation' (Miles and Snow, 1986; Reich, 1991), the 'shamrock organisation' (Handy, 1989), and the 'extended enterprise' (Browne *et al.*, 1995). Miles and Snow (1986) first used the term 'network organisation' which they argued had evolved as a result of rapid changes in product and process technologies and changes in the nature of international competition. Reich (1991) argued that outsourcing and the development of network organisations were essential if organisations were to compete in the new business environment. In his book, *The Age of Unreason*, Charles Handy (1989) used the term 'shamrock organisation', named after the three-leaf clover. One leaf represents the core organisation, the second flexible contract labour whilst the third leaf represents outsourced services and functions.

As well as being a major driver of changes in the business environment, rapid advances in ICTs have also acted as a key enabler in allowing organisations to pursue radically altered structures. Developments in ICTs have significantly increased both the complexity and amount of information that can be transmitted while at the same time reducing both the cost and time it takes. Furthermore, the geographical constraints associated with physical information transmission mechanisms have been effectively eliminated. These developments have led to a reduction in the costs of the dissemination of information both internally and externally for organisations. Prior to these developments, large hierarchical organisations were able to achieve significant cost advantages by generating, through their complex organisational structures, proprietary internal information storage and transfer capabilities that could not have been achieved outside a single organisation (Hendry, 1995). In these types of organisations, top-level managers had a wider and deeper span of knowledge and access to more information than employees at lower levels in the organisation. Cost and physical constraints limited the number of people within the organisation who could access information at the required level of detail. Furthermore, traditional organisation structure concepts such as span of control and hierarchical reporting led to employees being allowed access to the minimum amount of information. However, many of these organisations have redesigned their organisation structures and have used ICTs to co-ordinate and integrate the various elements of the business. Moreover, the advent of Internet technologies such as intranets and extranets has facilitated information exchange both vertically and horizontally in the organisation hierarchy as well as externally in the business network. For example, within large organisations, the emergence of open standards for exchanging information over Intranets fosters cross-functional teams and has the potential to accelerate the demise of hierarchical structures and their proprietary information systems (Evans and Wurster, 2000).

The adoption of new organisational structures has been a major driver behind the trend towards increased outsourcing. Competitive pressures and increased product complexity have led companies to create flatter, more flexible and responsive organisations in comparison with highly vertically integrated organisations. The emphasis on achieving greater efficiencies and lower costs has forced many organisations to increasingly focus in a limited number of core areas that drive competitive advantage. For example, Unilever, with a portfolio of 1600 food, toiletries and household products, decided that in order to increase sales and profitability, it would focus on a smaller number of 'power brands' – core products – that have global reach, thereby reducing costs and exploiting new distribution channels (Willman, 1999). In the past, organisations may have performed a range of activities internally based upon cultural, historical or political reasons rather than on the basis of enhancing the needs of customers or achieving competitive advantage. However, now many organisations have begun to challenge these assumptions and are restructuring their organisations in order to reflect changes in the business environment. Consequently, a major trend has been the use of outsourcing to access the capabilities of more competent external suppliers in a range of business activities.

2.5 The potential benefits of outsourcing

2.5.1 Cost reduction

Many organisations are motivated by cost considerations in adopting outsourcing strategies. In a study carried out by PriceWaterhouseCoopers (1999) it was found that most Western organisations primarily employed outsourcing to save on overheads through short-term cost reductions. For example, Chrysler estimated that in 1997, through increasing outsourcing of processes to suppliers, it would add $325 million to its annual profits and eventually generate over $1.2 billion in savings (Chalos and Sung, 1998). Outsourcing enables the customer to benefit from supplier cost advantages such as economies of scale, experience and location. Suppliers may take on investment and development costs while sharing these risks among many customers and thereby reducing supplier costs for all customers. For example, in the financial services industry many banks have outsourced high-volume transaction processing functions such as electronic payments and the processing of cheques to service providers with greater economies of scale in order to make the cost of each transaction much lower. In addition, by gradually outsourcing activities the customer can reduce risks by converting its fixed costs into variable costs. In times of adverse business conditions suppliers will then have to deal with the problem of excess capacity. However, suppliers should be better able to cope with demand fluctuations through economies of scale and have more scope for alternative sources for excess capacity.

2.5.2 Performance improvement

Many specialist suppliers can achieve much higher levels of performance in certain activities than can be achieved internally by the outsourcing company. This performance advantage is based not only in the form of reduced costs. Specialist suppliers can also provide the outsourcing company with a higher level of service quality. A key aspect of many outsourcing agreements is a service level agreement (SLA) that quantifies the service levels required by the outsourcing company. Maltz (1994) found in a study of warehousing outsourcing that the primary driver was an increase in the quality of service rather than a reduction in cost. In fact, a key reason why many organisations consider outsourcing is to improve perform-ance in an activity. In such circumstances, the outsourcing organisation must ensure that it has an effective measurement system in place to assess the performance level of suppliers on an ongoing basis.

2.5.3   Flexibility

In the past, many organisations have attempted to control the majority of activities internally on the assumption that controlling supply sources eliminates the possibility of short-run supply shortages or demand imbalances in product markets. However, such a strategy is both inflexible and inherently fraught with risks. Due to issues such as rapid changes in technology, reduced time-to-market and increasingly sophisticated consumers it is very difficult for organisations to control and excel at all the activities that create competitive advantage. Outsourcing can provide an organisation with greater flexibility, especially in the sourcing of rapidly developing new technologies or fashion goods. Specialist suppliers can provide greater responsiveness through new technologies than large vertically integrated organisations. In this way, with a network of specialist suppliers the customer organisation can rapidly increase or reduce production in response to changing market conditions at a lower cost.

2.5.4   Specialisation

Outsourcing can allow an organisation to concentrate on areas of the business that drive competitive advantage and outsource more peripheral activities enabling it to leverage the specialist skills of the supplier. Through extensive outsourcing organisations have created networks of product and service providers specialising in their own distinct area of expertise. For example, Quinn (1999) argues that specialists in supply markets can develop greater knowledge depth, invest more in software and training systems, be more efficient, and therefore offer higher salaries and attract more highly trained people than can the individual staff of all but a few integrated companies. These advantages can generate enough value to deliver a better service at a lower cost to the customer, whilst allowing the supplier to make a profit. Furthermore, specialisation can have a positive impact upon career development opportunities for employees. For example, in the case of employees in specialist functions such as information technology in large diverse organisations, the scope for career progression is normally limited to within a single function. However, there is evidence to show that when the same employees are transferred to a specialist organisation there are enhanced career and training opportunities. For example, Kessler et al. (1999) have found in a study of outsourcing in local government that employees were much more positive about career development opportunities after being transferred to the service provider. In a manufacturing context, this concept of specialisation has led organisations to seek competitive advantage in the activity of specifying and integrating external services and other purchases, rather than in assembly and production of the goods themselves. Many Western car makers have outsourced module assemblies such as chassis, climate-control systems and anti-lock braking systems to their suppliers. Many of these module suppliers locate in separate facilities near to the final assembly facility of the car maker. In some cases, these suppliers have established a direct connection with the final assembly lines within the car maker (Fredriksson, 2002).

2.5.5   Access to innovation

Many organisations are reluctant to outsource because they fear they may lose the capability for innovation in the future. However, in many supply markets there exists significant opportunities to leverage the capabilities of suppliers into the product and services of the customer organisation. Rather than attempt to replicate the capabilities of a supplier network it can be much more prudent to use outsourcing to fully exploit the suppliers' investments, innovations, and specialist capabilities. For example, suppliers provide almost all of Dell's component design, software and production requirements for its computers. It invests in areas where it perceives an opportunity for unique value added and avoids large inventory, facilities, and development risks incurred by many of its competitors (Quinn, 1999). Tesco, the food retailer, is pursuing a similar strategy in its efforts to offer retail telecommunications services (Nairn, 2003). Tesco has outsourced the entire operation to a number of service providers including Cable and Wireless for the telecommunications service, Servista for billing support and Vertex Customer Management for customer service. Each of these service providers is specialising in its own area of expertise whilst Tesco is enhancing its brand as a retail service provider. Interestingly, such skills at designing business systems and integrating the services and outputs of distinct units have become generally more desirable as companies move from strategies based on tangible assets and positioning to strategies based on intangible assets and capabilities (Grant, 1996). In a study of IT outsourcing Lacity *et al.* (1996) found that efficient supplier management practices could drive down costs more significantly than was achievable through economies of scale.

2.6   The risks associated with outsourcing

2.6.1   Cost increases

There is evidence to suggest that when organisations outsource to achieve cost reductions that costs do not decrease as expected and in some cases can increase.

When organisations outsource to achieve cost reductions, there is normally an early anticipation of cash benefits and long-term cost savings. However, many organisations fail to account for future costs and in particular that of managing the outsourcing process. For example, there is a tendency to under-estimate the management resources and time that have to be invested in outsourcing. Some organisations fail to realise that resources have to be invested in managing the relationship with the supplier, which is particularly important in the case of critical business activities. In many cases organisations outsource in order to effect improvements in certain parts of the business which have been causing problems. However, it is erroneous to assume that once the activity is outsourced the problem will disappear. For example, poor performance internally may have been due to weak management. It is often the case that the person previously responsible for managing the activity internally is responsible for managing the external supplier. This problem is further exacerbated if the outsourcing process has involved the transfer of staff from the outsourcing organisation to the supplier organisation. Therefore, it is important to determine why the activity is being outsourced in the first place. Also, when an organisation outsources it may not identify certain aspects of the activity that are provided internally. These can include issues such as employee goodwill or involvement of employees from other parts of the organisation (Lonsdale and Cox, 1997). With such aspects being omitted from the contractual agreement, the supplier may add on extra charges for these services.

2.6.2   Supply market risk

Organisations can encounter significant risks when they use supply markets for activities that they have performed internally in the past. Over dependency on a particular supplier can lead to significant risks in terms of cost, quality and supplier failure. For example, suppliers may fail to achieve the necessary quality standards demanded by the outsourcing organisation. Therefore, an organisation may decide to keep an activity in-house in order to guarantee quality and reliability of supply. In relation to the supply market, it is crucial for the outsourcing company to monitor changes in the supply market. When a company chooses to outsource it may do so on the basis of the presence of competition amongst a number of suppliers in the supply market. However, many organisations fail to monitor changes in the supply market and their impact upon the outsourcing process. For example, the balance of power may shift towards the supply market if there has been consolidation amongst a high number of small suppliers into a few large suppliers. Such a situation may arise if the company undertakes outsourcing with limited experience of managing a supplier relationship. Public sector organisations have had difficulties with monitoring and managing public sector contracts due to a lack of the requisite skills and experience. For example, in the US a report by the US General Accounting Office (1997) found performance management to be the most critical weakness in the privatisation initiatives pursued by the US government due to a lack of skilled personnel (cited in Kakabadse and Kakabadse, 2000). Many organisations fail to recognise that managing an external supplier requires a different set of skills than those associated with managing an internal process. Suppliers in outsourcing markets are extremely adept at exploiting any naiveté on the part of the outsourcing organisation in areas such as contract negotiation and relationship management.

2.6.3   Loss of skills

Outsourcing can lead to the loss of critical skills and the potential for innovation in the future. In the long-term an organisation needs to maintain innovative capacity in a number of key activities in order to exploit new opportunities in its respective markets. If an organisation has outsourced a number of critical activities its ability to innovate may be severely diminished. Innovation requires slack resources, organic and fluid organisational processes, and experimental competencies – all attributes that outside supply does not guarantee (Earl, 1996). These risks can become more pronounced when the objectives of the outsourcing company and the supplier are conflicting. For example, the customer organisation may decide to establish a short-term contract with a supplier in order to obtain the lowest price and keep the supplier in a weak position. However, this will seriously undermine any incentive for the supplier to pass on any of the benefits associated with innovation to the customer organisation. Suppliers can also become competitors in the future once they obtain the requisite knowledge to produce the entire product of which the outsourced activity is a significant element. An early example of this scenario is the Dodge Brothers, founders of what became a division of the Chrysler Corporation, supplying engines to the Ford Motor Company in the early 1900s. By 1914, they had vertically integrated forward to produce entire automobiles and were competing directly with Ford (Welch and Nayak, 1992). More recently, organisations from the Pacific Rim have also exploited the trend towards outsourcing by large Western manufacturers to develop the skills necessary to enter Western product markets and compete directly with companies that once were their customers.

2.6.4   Organisational change implications

Outsourcing has significant social implications for an organisation. For example, outsourcing can lead to the redeployment of staff within the customer organisation or the transfer of staff to the supplier organisation. The demands associated with outsourcing transcend organisational boundaries, and therefore the approach to managing the change process must ensure that complementary activities and behaviours are exhibited within and between organisations. However, organisations have had extreme difficulties with embracing and effectively managing the change process required. For example, a new focus on quality and customer relationships necessitates changes in policies, cultural values, work procedures and processes, relationship between departments, and interactions between buyers and suppliers. Often these elements of strategic change are not addressed by traditional strategic planning, which concerns itself with whether actions make financial and strategic sense (Worley *et al.*, 1996). Organisations often ignore the fact that successful outsourcing is heavily dependent upon the attitudes and commitment of their workforce. The perspectives and responses of employees at all levels and positions have a significant impact on the successful implementation of strategic change processes. Given the strategic nature of the decision to outsource, culture change is vital. However, effecting culture change is an enormous task. Frequently organisations fail to engage in a process whereby time, money and efforts are invested in bringing about a change in culture, structure and reward systems (Boddy *et al.*, 1998). The strategy of organisations tends to focus on content issues such as the achievement of efficiencies, while ignoring the process of how to achieve the efficiencies.

2.7   Inter-organisational relationship configurations

Organisations have been adopting a range of relationship configurations with suppliers and other organisations in order to reduce the risks associated with outsourcing. In particular, organisations that have adopted extensive outsourcing strategies have attempted to adopt collaborative arrangements with their key suppliers. The relationship configurations adopted have been influenced by the type of product or service being outsourced and the number of capable suppliers that can deliver the product or service. In the case of a standard product or service that can be supplied by a number of external providers such as catering or security, the outsourcing organisation is likely to employ a relationship bounded by explicit contractual safeguards such as price and payment terms, a short-term perspective and a clear definition of roles and responsibilities. Alternatively, in the case of a more critical product or service, although a contract will be established at the outset, the outsourcing organisation is also likely to build a more collaborative relationship through relational mechanisms such as bi-directional information sharing, longer term perspective and joint problem solving. These collaborative arrangements are sometimes referred to as 'quasi-integration' arrangements and can include strategies such as joint ventures, strategic alliances, franchising and partnership sourcing. It is important to stress that many of these configurations have been employed before the trend towards increased outsourcing. However, the increasing externalisation of supply has forced many organisations to pursue more innovative relationship configurations with their suppliers. Some of these relationship configurations are now discussed.

2.7.1   Joint ventures and strategic alliances

Joint ventures and strategic alliances allow organisations to exchange certain products, services, information, and expertise while maintaining a formal trading relationship on other parts of the business. For example, many manufacturing companies are increasingly relying upon competent suppliers, who have been able to contribute to the customer's efficiency in production as well as research and development. In other words, these organisations have been supplementing their key capabilities by allying with other providers of complementary capabilities to satisfy the needs of their customers. Virgin, which encompasses retailing, travel, soft drinks and music, provides a good example of a joint venture, where one party provides the brand name and marketing expertise (Virgin), and other parties provide the production facilities and capital (Johnson and Scholes, 2002). Strategic alliances are essentially agreements between two or more companies to share the costs, risks and benefits associated with exploiting new business opportunities. An alliance can take the form of a long-term contract between companies in which they agree to undertake some joint activity that benefits both. For example, Kodak established strategic alliances by awarding 5- and 10-year contracts with a number of information technology service providers including IBM (McFarlan and Nolan, 1995). The agreements were structured to allow the suppliers to make a profit and to encourage them to continuously improve mutually beneficial areas. Creating spin-off companies has been another 'quasi-integration' arrangement. Companies may develop world-class capabilities in non-critical activities that may spin-off into separate profit-making entities. For example, Daimler–Chrysler developed world-class capabilities in supporting activities such as information technology, finance and telecommunications. In the area of information technology, Daimler–Chrysler created a service company, Daimler–Chrysler Services in order to transform its IT capabilities into a profitable business. Similar arrangements are also prevalent at the customer side of the business. Franchising allows the lead organisation to control the distribution of the product or service without having to invest resources in the physical infrastructure. Such an arrangement allows the lead organisation to avoid ownership of the physical assets but retain ownership of the intangible assets such as brand identity and corporate image.

2.7.2   Collaborative buyer–supplier relationships

Perhaps the most prominent development in inter-organisational configurations has been the adoption of collaborative buyer–supplier relationships. The type of collaborative arrangement being adopted by many Western organisations has been heavily influenced by the relationships pursued by Japanese companies and their key suppliers. Japanese companies have frequently engaged in 'partnership' buyer–supplier relationships, with buyers utilising target pricing rather than competitive bidding. In their book *The Machine that Changed the World*, Womack *et al.* (1990) explained how companies could dramatically improve their performance by adopting the 'lean production' approach pioneered by Toyota. By eliminating unnecessary steps, aligning all stages in an activity in a continuous flow, combining employees into cross-functional teams and continually striving for improvement, companies can develop, produce and distribute products with half or less of the human effort, space, tools, time and overall expense. The term 'lean' embodies a system that uses less of all inputs to create outputs similar to the mass production system but offering an increased choice to the end customer. However, a significant part of the success of the Toyota Production System (TPS) can be attributed to the integration of their key internal processes with those of their suppliers (Hines, 1996). The objective is to create a win–win situation for the customer and supplier partners in order to reduce costs and lead times. For example, the need for minimal inventory for cost and quality reasons and early detection of defects requires frequent deliveries from suppliers. Suppliers are required to deliver frequently, in small quantities, as required to the point of use with total quality guaranteed eliminating the need for incoming inspection. Suppliers are also involved in the design of components with assemblers, organising their supply base into a tiered hierarchical structure. The notion here is that buyers and suppliers are locked together in long-term, obligational relations as distinct from the historical Western model of short-term competitive relations where contracts are awarded largely on the basis of price. The supply base plays an important role in the larger value-adding network of the buyer organisation by facilitating the adding of value in the delivery of the product to the final user. Furthermore, this contrasts with the mass production paradigm where suppliers were only concerned with the next customer in the supply chain and were largely oblivious to the needs of the final customer of the product.

The Japanese model of sub-contracting reinforces the key principles that are intrinsic to the culture of the country. The development of long-term partnerships is based on the acceptance of the hierarchical nature of Japanese business and its implications for power exchanges between the larger and smaller companies which exist (Barnett *et al.*, 1995). Supplier partners are typically called *kankei-gaisha* (affiliated companies) in Japan and are considered to be the vertical *keiretsu* of the parent company. It is not unusual for the parent company to hold a stake in these affiliated companies. For example, NEC directly owns substantial shares of stock in the companies that supply it with parts, designs, components, subassemblies and services (Harrison, 1994). The enormous benefit of the Japanese tradition has been the ability of large companies to focus on the needs of the entire value system unimpeded by functional concerns, career paths within functions, and the constant struggles between members of the value system to gain advantage over each other (Womack and Jones, 1996). Under the leadership of the larger companies, the members of the *Keiretsu* are always stimulating each other to achieve competitiveness and to enhance productivity further through such means as technology exchange, shared responsibility for technology development and other mutual support activities. However, it must be emphasised that this is not a 'relaxed and cosy' relationship between the buyer and supplier. Womack *et al.* (1990) have found that Japanese suppliers face constant pressure to improve their performance, both through constant comparison with other suppliers and contracts based on reduced costs and delivery prices. Although there has been much debate on the applicability of the principles associated with the Japanese sub-contracting model in a Western context, it has been extremely influential on the development and practice of supply chain management and in particular collaborative buyer–supplier relationships. Indeed, the dismantling of a number of *keiretsus* in Japan as a result of the financial crisis in Asia in the 1990s has cast doubts on the applicability of some of the features associated with Japanese style partnerships.

2.7.3   Virtual corporations

A further development in this area has been the development of virtual corporations. A virtual corporation is an organisation composed of a number of business partners sharing costs and resources in the development, production and delivery of a product or service. The philosophy of the virtual corporation is to create a high level of trust and eliminate the potential for opportunism through making the organisations involved more dependent on one another. Moreover, each organisation involved in the virtual corporation is the most competent available in its specialist area. In effect, the virtual corporation is a competence-based network of best-in-class product and service providers carrying out a sequence of value-adding activities in the value chain. Instead of directly owning assets or employing staff, it relies heavily on 'getting things done' rather than doing things itself. For example, some airline companies have become increasingly virtual, leasing their planes and buying in engineering and maintenance services, catering, and ground support services. Virtual corporations offer the following advantages (Alexander, 1997).

- *Focus* – by focusing on a smaller number of activities than competitors, the virtual corporation can overcome resource constraints and develop competencies in areas that are likely to create competitive advantage.
- *Access* – the virtual corporation can gain access to best-in-class providers of products and services that in turn enhance the quality and performance of its end products.
- *Flexibility* – in rapidly changing environments it is more beneficial to have access to sources of supply rather than own the physical assets associated supply sources.

Virtual corporations can be permanent-type arrangements designed to create or assemble production resources rapidly, frequently, concurrently or to create or assemble a broad of productive resources (Goldman *et al.*, 1995). Furthermore, a virtual corporation can be established for a specific project. The creation and operation of a virtual corporation depends very much upon the utilisation of information technologies to enable rapid communication and collaboration amongst each business partner involved. In particular, developments in Internet technologies such as extranets facilitate the operation of the virtual corporation. Extranets are networks that link one organisation with another over the Internet. Extranets can provide secured connectivity between an organisation's Intranets and the Intranets of its business partners, materials suppliers, financial services, government and customers. The protected environment of the extranet allows groups to collaborate, share information exclusively and exchange it securely.

Illustration 2.2

The SMART arrangement

An interesting illustration of a virtual corporation is the arrangement adopted for the Smart Minicar, the unconventional two-seater small city car developed by Micro Compact Car (MCC). This arrangement involved a joint venture between Mercedes–Benz and SMH (the Swiss company that makes the range of 'Swatch' products). The initial impetus for the project came from Nicolas Hayek, the CEO of SMH. He viewed the project as a vehicle for further exploiting the Swatch concept and company capabilities in marketing and distribution. He also believed that SMH's capabilities in designing and building microelectronic propulsion systems could be applied in the automotive industry as well contributing to the development of environmentally friendly propulsion systems with lower fuel consumption levels. In order to access the automotive expertise required for such an arrangement, he approached Mercedes–Benz who were willing to become involved in project. Based upon market research the partners established a vision for the project as that of developing a city car that was small in size whilst maintaining passenger comfort and safety, low fuel consumption, the use of recyclable materials, and involved an environmentally friendly production process. In order to realise this vision there were a number of technological innovations achieved including the use of lightweight materials in the engine, lightweight body panels to improve fuel consumption, a 50% component reduction in the design of the engine turbocharger. The Tridion-frame technology invented by Mercedes–Benz was used as a steel-faced body for the car around which the entire vehicle was designed. Also, applying some of the innovative marketing approaches associated with Swatch, the customer was allowed to choose the colours of the body panels and rapidly alter these colours in the dealership after purchase if necessary. Indeed, the arrangement received funding from the European Union in recognition of its environmentally friendly production system, developing a new market segment and the innovativeness of the concept.

The joint venture was based on extremely limited resources and a relatively short development to full production timescale of around 4 years. Due to restraints on finance and human resources, the companies created an organisational structure that involved MCC co-ordinating a group of key suppliers that would provided more than 85% of the value-added of the 'Smart'. The design of the car facilitated the use of this approach. The architecture of the design of the car was segmented into a number of modules and systems. A module was defined as a major sub-system of the car such as the cockpit, the doors or drive train. Each module contained a number of systems such as the air management system, the brake system or the wiper system. The development and production of these entire modules and systems were then outsourced to one of the key suppliers. Suppliers played an unprecedented role in this arrangement. For example, companies such as Bosch (front modules), Dynamit Nobel (plastic body panels) and Krupp–Hoesch (mechanical sub-assembly, including the motor and axles), were involved in the project from the design stage. Their roles involved consultation on the basic design of the vehicle to the co-ordination of development and production engineering and even direct involvement in the layout and workflow in the factory. The companies involved pre-assembling the important modules of the vehicle within the plant or linked facilities in a nearby industrial park – referred to as 'Smart Ville' by MCC. Steel came from Magna International, the Canadian group. VDO, the German automotive electronics maker built fully assembled cockpits, including the instruments while another German components specialist, Ymos, made complete door assemblies, including the trim, control system for the windows and glass.

The success of this arrangement depended heavily upon the level of cooperation and agreement about the division of roles and responsibilities between MCC and the key suppliers. Process definitions were developed to clarify mutual targets and allow for co-ordination and changes between MCC and the suppliers. For example, the relationships between MCC and the key suppliers were based on contracts and rules that each party had to agree. Contracts were based on single sourcing arrangements with each supplier expected to take on some of the business risk associated with the project by providing finance on system and module development. MCC attempted to create transparency in the relationship between the key suppliers in order to reduce conflict and build trust. For example, they operated 'open-book calculations' with suppliers that involved the sharing of information that would enhance the relationship. Suppliers were guaranteed purchasing volumes from MCC and access to estimated product sales volumes. The design and layout of the site was also structured in order to facilitate greater co-operation and communication. The main assembly area was a large building in the shape of a mathematical plus sign. Around it were the separate premises of the system partners linked by conveyor belts. This is shown in Figure 2.2. Shaping the assembly area like a 'plus' allowed the system suppliers to be located at just the right place on the assembly line for the installation of the modules. Employees from the system partners could mix freely and were treated identically to MCC's own staff and take charge of final assembly within the 'plus'.

Sources: Simonian, H. (1998). Smart work for partners. *Financial Times – Automotive Industry Review: London*, 23 February, p. 7.
Pfaffmann, E. and Bensaou, B. M. (2004). Mercedes–Benz and Swatch: inventing the 'smart' and the networked organisation. Case Study 33 in *Strategic Management and Business Policy*, Wheelen, T. L. and Hunger, J. D. (eds.), ninth edition. Upper Saddle River, New Jersey: Pearson Prentice-Hall.

2.7.4   Public sector relationship configurations

Public sector organisations have also adopted a range of relationship configurations with external parties in order to reduce the risks associated with the outsourcing of products and services. The public–private partnership (PPP) philosophy has attempted to create an environment of co-operation between the public and private sectors in the form of an inter-organisational partnership. PPPs not only involve the introduction of market mechanisms to public sector management but also involve using partnerships to integrate the strengths of public and private sectors in order to achieve a common set of objectives. Public–private partnerships can be traced back to the 1960s when partnerships were employed by the federal government in the US as a means of encouraging investment from the private sector in the development of inner-city

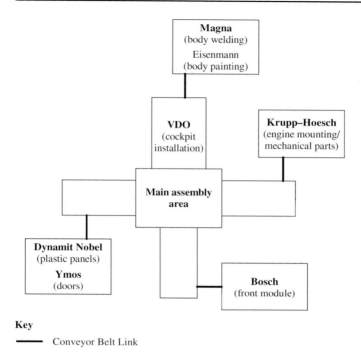

Key

———— Conveyor Belt Link

Figure 2.2 Smart Ville's integrated suppliers (adapted from Simonian, 1998)

infrastructures. Ideally, public and private sector organisations are both pursuing mutual advantages in creating a partnership that is characterised by trust, openness, fairness and mutual respect. The objectives of the public sector organisation in such an arrangement include partnering with the private sector in order to improve performance, reduce costs, obtain higher service levels, and share risks and responsibilities. The objectives of the private sector organisation include an attractive investment proposition, to achieve a profit, and an opportunity to further develop its business. The concept of the public–private partnership involves a number of key elements including acting in and protecting the public interest and the stimulating of investment from the private sector (Pongsiri, 2002). For example, a major element of public–private partnerships in the UK has been the public finance initiative (PFI) which involves the purchase of an asset by the public sector being replaced by the purchase of a service for an annual fee being paid to the contractor (Akintoye *et al.*, 1998). The structure of a PFI project involves a number of elements including a concession agreement, a contract between a public sector organisation that grants a promoter a right or privilege to undertake construction works, along with an obligation to undertake those works for a specified time period. As with partnership arrangements in any context their success depends upon both parties having mutual and compatible objectives.

In the context of PPPs there have been difficulties with reconciling the goals of public and private sector organisations. Private sector organisations are primarily driven by the goal of profit maximisation whilst public sector organisations have to pursue goals that are in the public interest. Moreover, there is a greater onus on private sector organisations to consider their social responsibility. In fact, experience with PPPs has shown that regulation is required in order to strike a balance between the interests of both the public and private sectors. Savas (2000) has argued that some form of regulation should be employed in order to ensure competition and adherence to market mechanisms without exercising unnecessary controls. There is also the problem of loss of control once market mechanisms are introduced to a previously hierarchical control structure. For example, the privatisation of the British Railway network led to a breakdown in control that many blamed on privatisation. Privatisation led to the previous hierarchical structure being restructured into over more than 100 service providers bound together through a range of complicated mechanisms including legal contracts, performance measures and penalty payments. However, the complexity of these arrangements made it increasingly difficult for management to control the railway system. Indeed, the company decided to bring railway maintenance back in-house. When the rail network was privatised in the UK, track maintenance was outsourced to a number of private engineering contractors. However, a fatal accident and a number of derailments had led to considerable criticism being directed towards both Network Rail and its contractors including Jarvis. One of the results of this was that Jarvis, the private contractor responsible for track maintenance decided that it no longer wished to undertake the work. As a result, Network Rail decided to transfer 18 000 staff from the private engineering companies back in-house in order to focus on rebuilding and maintaining the railway network.

Some of the discussion on the influence of collaborative or partnership arrangements on inter-organisational relationships has highlighted the dangers of conflicting objectives between each party to the relationship. These conflicting objectives between organisations and their suppliers are also consistent with some of the potential difficulties that can arise between organisations and their employees. Many organisations are often unwilling to consider outsourcing because they fear being held to ransom by an opportunistic supplier in the supply market. However, these same organisations fail to recognise each time they recruit an employee from the labour market they are incurring some level of risk. In fact, for many organisations the most opportunistic party that they are ever likely to establish a relationship with is their employees. Organisations erroneously make the assumption that once they recruit an employee that they can exercise considerable control over that employee. In effect, they are equating ownership with control. However, in many organisations senior management exercise limited control over many of their employees. For example, many public sector organisations with rigid bureaucratic structures employ individuals over whom they have little control. Attempts to rectify such anomalies are seriously constrained by the presence of rigid and inflexible contracts and powerful unions. In many instances the goals of the employee and organisation are not aligned. Indeed, many organisations outsource activities to suppliers because they feel they can exercise more control over an external supplier than an internal business function. For example, Clark (1989) found in interviews with project managers and engineers in manufacturing companies that in some instances they were in a better position to exercise greater control over external suppliers than their internal parts division. Also, in environments such as information technology and electronics where employees are increasingly mobile and there is a low level of job security with a single organisation, employees no longer 'belong' to a single organisation. This trend has become increasingly pronounced as individuals develop their careers in a way that means they belong to a profession or specialism rather than a single organisation.

Illustration 2.3

Outsourcing and the consequences of failure
There have been a number of high profile cases of outsourcing failures. These failure cases have been particularly well documented in the public sector due to the high level of scrutiny and accountability associated with the delivery of public services. One such case in the UK has been attempts, through a PFI, by the Lord Chancellor's Department to fully computerise and integrate the nation's magistrates courts – known as the Libra project. Traditionally, case files on defendants, convicted or on remand, have been distributed by fax or post. The National Audit Office in the UK found that over half of ineffective hearings were as a result of poor liaison within or between courts and other agencies. The Libra project was viewed as a flagship government project to speed up the criminal justice system by providing advanced electronic links between magistrates courts and other agencies. The initial objective of the project was to streamline more than 300 magistrates court procedures and allow criminal cases to be processed much more rapidly between the police, prisons, probation services and other agencies. However, the project failed to deliver upon its initial objectives. Indeed, at one stage the project was on the brink of being abandoned. The Public Accounts Committee condemned the Libra project as one of the worst PFIs describing it as 'disastrous at every turn'. For example, the cost of the project rose from an initial estimate of £146 million to around £400 million, whilst failing to deliver a fully operational system. It was found in some cases that staff in many courts were working with two PCs on their desk. They were using the new PC for basic tasks such as word

processing whilst using their old PC, installed 10 years previously, to access casework.

The failure to meet its objectives and the cost overruns provoked much controversy from both the public and politicians alike. The project involved the installation of 11 000 PCs with the installation cost for each computer terminal in each court estimated at around £21 000. One public representative remarked upon hearing this finding that it was possible to purchase a personal computer in the high street for around £700. However, if one considers the way in which the project was managed, it is not difficult to understand why there was a failure to deliver on the initial objectives. The department responsible had mishandled the tendering process for the project. The tendering process only attracted one bidder – ICL (later taken over by Fujitsu). All the potential bidders except ICL dropped out at the procurement stage. In fact, the lack of potential bidders at the procurement stage should have alerted the department that the project may not have been sufficiently well designed to attract competition. At the outset ICL was not clear on what the requirements of the department were in relation to the overall project. ICL submitted an under-priced bid and failed to meet key delivery dates. The government department, in turn, failed to take decisive action when ICL failed to deliver upon the initial requirements. Initially, ICL was awarded a 10-year contract costing £146 million. However, as a result of the difficulties associated with the project the contract was amended in 2003. The UK government agreed to pay £390 million to Fujitsu and additional suppliers for an 8-year contract that was regarded as not as comprehensive as the original contract.

Sources: Hencke, D. (2003). Court computer plan wastes millions – and still doesn't work. *The Guardian*, London, 11 November, p. 1.
Millar, S. (2002). Humiliation looms over IT system for courts. *The Guardian*, London, 25 April, p. 13.

REFERENCES

Akintoye, A., Taylor, C. and Fitzgerald, E. (1998). Risk analysis and management of private initiative projects. *Engineering Construction and Architectural Management*, **5**, No. 1, 9–21.

Alexander, M. (1997). Getting to grips with the virtual organisation. *Long Range Planning*, **30**, No. 1, 122–4.

Anonymous, (1997). Outsourcing of HR continues. *HR Focus*, **74**, No. 3, 2.

Barnett, H., Hibbert, R., Curtiss, A. and Sculthorpe-Pike, M. (1995). The Japanese system of subcontracting. *Purchasing and Supply Management*, December, 22–26.

Barthelemy, J. (2001). The hidden costs of IT outsourcing. *Sloan Management Review*, **42**, No. 3, 60–9.

Barthelemy, J. and Geyer, D. (2000). IT outsourcing: findings from an empirical survey in France and Germany. *European Management Journal*, **19**, No. 2, 195–202.

Bettis, R. A. and Hitt, M. A. (1995). The new competitive landscape. *Strategic Management Journal*, **16**, 7–19.

Boddy, D., Cahill, C., Charles, M., Fraser-Jraus, H. and MacBeth, D. (1998). Success and failure in implementing supply chain partnering: an empirical study. *European Journal of Purchasing and Supply Chain Management*, **2**, No. 2/3, 143–51.

Browne, J., Sackett, P. J. and Wortmann, J. C. (1995). Future manufacturing systems – towards the extended enterprise. *Computers in Industry*, **25**, 235–54.

Chalos, P. and Sung, J. (1998). Outsourcing decision and managerial incentives. *Decision Science*, **29**, No. 4, 901–19.

Christopher, M. (1998). *Logistics and Supply Chain Management: Strategies for Reducing Cost and Improving Service*. London: Financial Times: Prentice Hall, second edition.

Clark, K. B. (1989). Project scope and project performance: the effects of parts strategy and supplier involvement on product development. *Management Science*, **35**, No. 10, 1247–63.

Cook, P. (1999). Privatisation and utility regulation in developing countries: the lessons so far. *Annals of Public and Co-operative Economics*, **70**, No. 4, 549–87.

Earl, M. J. (1996). The risks of outsourcing IT. *Sloan Management Review*, **37**, No. 3, 26–32.

Domberger, S. (1998). *The Contracting Organisation: A Strategic Guide to Outsourcing*. Oxford: Oxford University Press.

Dyer, J. H. and Ouchi, W. G. (1993). Japanese-style partnerships: giving companies a competitive edge. *Sloan Management Review*, **34**, No. 1, 51–63.

Evans, P. B. and Wurster, T. S. (2000). *Blown to Bits: How the New Economics of Information Transforms Strategy*. Boston: Harvard Business School Press.

Fredriksson, P. (2002). Modular assembly in the car industry – an analysis of organisational forms' influence on performance. *European Journal of Purchasing and Supply Management*, **8**, 221–33.

Goldman, S., Nagel, R. and Preiss, K. (1995). *Agile Competitors and Virtual Organisations*. New York: Van Nostrand Reinhold.

Grant, R. M. (1996). Toward a knowledge-based theory of the firm. *Strategic Management Journal*, **17**, Winter special issue, 109–22.

Handy, C. (1989). *The Age of Unreason*. Boston: Harvard Business School Press.

Harrison, B. (1994). *Lean and Mean: The Changing Landscape of Corporate Power in the Age of Flexibility*. New York: Basic Books.

Hendry, J. (1995). Culture, community, and networks: the hidden cost of outsourcing. *European Management Journal*, **13**, No. 2, 193–200.

Hill, C. W. L. and Jones, G. R. (2001). *Strategic Management: An Integrated Approach*. New York: Houghton Mifflin Company, fifth edition.

Hines, P. (1996). Purchasing for lean production: the new strategic agenda. *International Journal of Purchasing and Materials Management*, **32**, No. 4, 2–10.

Johnson, G. and Scholes, K. (2002). *Exploring Corporate Strategy*. London: Financial Times: Prentice Hall, sixth edition.

Kakabadse, N. and Kakabadse, A. (2000). Critical review – outsourcing: a paradigm shift. *The Journal of Management Development*, **19**, No. 8, 670–728.

Kessler, I., Coyle-Shapiro, J. and Purcell, J. (1999). Outsourcing and the employee perspective. *Human Resource Management Journal*, **9**, No. 2, 5–19.

Kutnick, D. (1999). The externalisation imperative. *CIO: The Magazine of for Information Executives*, February, 27–29.

Lacity, M. C., Willcocks, L. P. and Feeny, D. F. (1996). The value of selective IT outsourcing. *Sloan Management Review*, **37**, 13–25.

Liker, J. K. and Wu, Y. C. (2000). Japanese automakers, US suppliers and supply chain superiority. *Sloan Management Review*, **41**, No. 1, 81–93.

Lonsdale, C. and Cox, A. (1997). Outsourcing: risks and rewards. *Supply Management*, 3 July, 32–4.

Lonsdale, C. and Cox, A. (2000). The historical development of outsourcing: the latest fad? *Industrial Management and Data Systems*, **10**, No. 9, 444–50.

Maltz, A. B. (1994). The relative performance of cost and quality in the outsourcing of warehousing. *Journal of Business Logistics*, **15**, 45–62.

Mathiason, N. (2004). Fair trade: coming to a shelf near you. *The Observer – Business Supplement: London*, 29 February, p. 7.

McFarlan, F. W. and Nolan, R. L. (1995). How to manage an IT outsourcing alliance. *Sloan Management Review*, **36**, No. 2, 9–23.

Miles, R. and Snow, C. (1986). Network organisations: new concepts for new forms. *California Management Review*, **28**, No. 3, 62–73.

Nairn, G. (2003). A new chapter for outsourcing. In: *FT Understanding Business Agility Part 2*. London: Financial Times, pp. 4–5.

Osborne, D. and Gaebler, T. (1992). *Reinventing Government: How the Entrepreneurial Spirit is Transforming the Public Sector*. Reading. MA: Addison-Wesley.

Picot, A., Reichwald, R. and Wigand, R. (1997). *Information Organisation and Management: Expanding Markets and Boundaries*. New York: Wiley.

Pongsiri, N. (2002). Regulation and public–private partnerships. *International Journal of Public Sector Management*, **15**, No. 6, 487–95.

Porter, M. E. (2001). Strategy and the Internet. *Harvard Business Review*, **79**, No. 2, 63–78.

PriceWaterhouseCoopers (1999). *Global Top Decision Makers' Study on Business Process Outsourcing*. New York: PriceWaterhouseCoopers, Yankelovich Partners, Goldstain Consulting Group.

Quelin, B. and Duhamel, F. (2003). Bringing together strategic outsourcing and corporate strategy: outsourcing motives and risks. *European Management Journal*, **21**, No. 5, 647–61.

Quinn, J. B. (1999). Strategic outsourcing: leveraging knowledge capabilities. *Sloan Management Review*, **40**, No. 4, 9–21.

Reich, R. (1991). *The Work of Nations*. New York: Simon and Schuster.

Sampler, J. L. (1998). Redefining industry structure in the information age. *Strategic Management Journal*, **19**, 345–55.

Saunders, C., Gebelt, M. and Hu, Q. (1997). Achieving success in information systems outsourcing. *California Management Review*, **39**, 63–79.

Savas, E. S. (2000). *Privatisation and Public–Private Partnerships*. New York: Chatham House Publishers.

Stiglitz, J. (2002). *Globalisation and its Discontents*. London: The Penguin Press.

Treacy, M. and Wiersema, F. (1993). Customer intimacy and other value disciplines. *Harvard Business Review*, **71**, 84–93.

US General Accounting Office (1997). *Privatisation: Lessons Learned by State and Local Governments*, GAO-GGD 97–48. Washington DC: US General Accounting Office.

Welch, J. A. and Nayak, P. R. (1992). Strategic sourcing: a progressive approach to the make or buy decision. *Academy of Management Executive*, **61**, 23–30.

Willman, J. (1999). Unilever to focus on core 'power brands'. *Financial Times, London*, 22 September, p. 25.

Wirtz, B. W. (2001). Reconfiguration of value chains in converging media and communications markets. *Long Range Planning*, **34**, 489–506.

Womack, J. P. and Jones, D. T. (1996). *Lean Thinking*. New York: Simon and Schuster.

Womack, J., Jones, D. T. and Roos, D. (1990). *The Machine that Changed the World*. New York: Rawson Associates.

Woodall, P. (2000). Untangling E-conomics. *The Economist – A Survey of the New Economy, London*, 23–29 September, pp. 5–9.

Worley, C., Hitchen, D. and Ross, W. (1996). *Integrated Strategic Change: How OD Builds Competitive Advantage*. Reading, MA: Addison-Wesley.

3 Theoretical influences on outsourcing

3.1 Introduction

This chapter provides an overview of the theoretical perspectives that have influenced the development of the framework for outsourcing evaluation and management. The framework draws upon a wide body of literature from a number of areas including business strategy, economics and inter-organisational relationships. This chapter outlines the characteristics and limitations of a number of theories from these areas including transaction cost economics (TCE), the resource-based view of the firm (RBV), the industry view (IV) and the relational view (RV). The analysis of the literature in these areas has had a major influence on the development of the outsourcing framework, which includes a number of aspects from each theoretical standpoint. It is argued that there are a number of inter-dependencies with each of the theoretical perspectives that can assist in outsourcing evaluation and management. The framework recognises the importance of the business strategy and inter-organisational relationship literature to an understanding of outsourcing. There is a growing emphasis in the business strategy literature of the importance of managing beyond the boundaries of the firm as well as within them. This chapter serves as a preface to the following chapter, which provides an outline of the practical problems with outsourcing and an overview of the stages in outsourcing evaluation and management.

3.2 Transaction cost economics

The most influential theory on outsourcing has been Williamson's theory of transaction cost economics (1975). Transaction costs analysis combines economic theory with management theory to determine the best type of relationship a firm should develop in the market place. This has laid the foundations for the purchasing discipline that uses an analysis of the factors that determine the internal and external boundaries of the firm. The concept of transaction cost analysis is that the properties of a transaction determine what constitute the efficient governance structure – market, hierarchy or alliance. The primary factors producing transactional difficulties include the following.

- *Bounded rationality* – the rationality of human behaviour is limited by the ability of the actor to process information.
- *Opportunism* – people are prone to behave opportunistically which leads to self-interest seeking with guile.
- *Small numbers bargaining* – many bargaining situations are infrequent or involve small quantities where the cost of obtaining full information is prohibitive; i.e. as in an oligopoly.
- *Information impactedness* – asymmetrical distribution of information among the exchanging parties that means that one party might have more knowledge than another.

These transaction difficulties and associated costs increase when transactions are characterised by (Williamson, 1985):

- *Asset specificity* – transactions which require high investments which are specific to the requirements of a particular exchange relationship.
- *Uncertainty* – ambiguity as to transaction definition and performance.
- *Infrequency* – transactions which are seldom undertaken.

The central theme of transaction costs theory is that the properties of the transaction determine the governance structure. Asset specificity refers to the non-trivial investment in transaction-specific assets (Williamson, 1985). The level of customised equipment or materials involved in the transaction between the buyer and supplier relates to the degree of asset specificity. When asset specificity and uncertainty is low, and transactions are relatively frequent, transactions will be governed by markets. Governance through markets can be described as discrete contracts that are short-term, bargaining relationships between highly autonomous buyers and suppliers designed to facilitate an economically efficient transfer of property rights (Ring and van de Ven, 1992). For example, manufacturing firms employ this governance structure in the case of standardised components that can be procured from a number of suppliers. High asset specificity and uncertainty lead to transactional difficulties with the transaction held internally within the firm – hierarchical governance. For example, a retail organisation is employing hierarchical governance when it owns and manages it store network. Medium levels of asset specificity lead to bilateral relations in the form of co-operative alliances between the organisations – intermediate governance. For example, two organisations may form a strategic alliance to jointly design and market a new product. Therefore, there is a degree of dependency that one or both organisations in the co-operative alliance can take advantage of. Collaborative arrangements attempt to pre-empt opportunism, by recognising the opportunity to work together for mutual benefit, in a long-term on-going relationship. The two extremes of the sourcing decision are either vertical integration or outsourcing.

The key issue in the sourcing decision is determining the boundaries between these two extremes. Williamson (1985) argues that the decision will always be made in relation to the scope for cost reduction and the importance of asset specificity. Therefore, the company should outsource activities if to carry them out internally would require excessive investment to get the lowest unit cost.

When using the external supply market, the TCE approach proposes the employment of contractual safeguards to counter any potential risks that might arise in the buyer–supplier relationship. In the TCE approach, the buyer must establish a governance arrangement with the supplier in order to ensure the delivery of a product or service at a specified price, quantity and quality level. More complex contractual safeguards are required as the level of risk in the relationship increases. The TCE approach proposes the use of contractual safeguards in the following situations (Poppo and Zenger, 2002).

- *Asset specificity* – high levels of asset specificity in the relationship, for example through investments in equipment, require the employment of contractual safeguards. For example, a supplier may have to develop a customised product or service for the buyer. The buyer, in turn, may need to develop an understanding of the supplier's processes in order to fully exploit its capabilities. Maintaining and developing the relationship is essential to realising the value of the investments made by each organisation in the relationship. However, in order to limit the threat of either organisation attempting to behave opportunistically, contractual safeguards should be introduced. These contractual safeguards should promote a long-term relationship through stipulating required actions and conditions in the case of a breach of contract and a mechanism for resolving disputes.
- *Performance measurement* – the presence of performance measurement difficulties increases the need for contractual safeguards. Measurement difficulties create the potential for each party to renege on their requirements. In order to reduce the likelihood of this situation, more complex contracts can be designed which allow more accurate measurement mechanisms. For example, the contract can include clauses to allow third-party performance monitoring and benchmarks to assess performance.
- *Uncertainty* – creates difficulties in the relationship because either the buyer or supplier may have to adapt in response to unexpected changes. For example, rapid advances in technology can create a high level of uncertainty in the relationship. Furthermore, a supplier may not wish to make a specific investment in the relationship without the necessary contractual safeguards. In this case, straightforward market governance is not sufficient to allow for this uncertainty. More complex contracts can specify clauses and procedures that allow for negotiations in the case of unexpected changes.

The fundamental question associated with the TCE approach is: when does the firm use market governance, hierarchical governance or intermediate governance?

The answer to this question is determined by both the cost of governance and the threat of opportunism. The assumption of the TCE approach is that companies will be motivated primarily by efficiency considerations and choose the least costly of the options. For example, the cost of using market governance is less than the cost of both intermediate and hierarchical governance. However, the potential for opportunism associated with each governance option must be considered. The potential for opportunism exists when one party in the relationship has made a transaction-specific investment, whilst the other party has not made a similar investment. In this case, the party that has made the transaction-specific investment is prone to opportunistic behaviour from the party that has not made the investment. The TCE approach argues that the organisation can use the appropriate governance mechanism to reduce the threat of opportunism. For example, when the exchange is characterised by high levels of transaction-specific investments, the threat of opportunism can be reduced through employing hierarchical governance.

However, in some contexts this analysis is inappropriate. For example, an organisation may require a capability in order to compete effectively in a particular market. According to the TCE approach, if the level of transaction-specific investment is high then the organisation should employ hierarchical governance. This will involve either developing the capability internally or acquiring an organisation that possesses the required capability. However, it may be extremely costly for an organisation to pursue either of these options. Therefore, the decision does not depend solely on the level of transaction-specific investment but also on the costs of internal development or acquisition. In the case of significant transaction costs, the organisation may have no choice other than to pursue other forms of governance even if the threat of opportunism exists. Barney (1999) argues that the opportunism that arises from transaction-specific investment is the cost of accessing capabilities that are too costly to obtain in other ways; i.e. hierarchical governance. This analysis is particularly pertinent to high technology industries in which organisations have to rapidly access capabilities that are difficult to replicate.

The TCE approach provides a sound theoretical basis for understanding market versus hierarchical governance mechanisms for determining the boundary of the firm. However, Doz and Prahalad (1991) argue that transaction cost analysis is limited because it focuses on single transactions as the unit of analysis. The emergence of collaborative arrangements in many industries involves repeated transactions between the same partners. The emphasis on market and hierarchical governance is a significant weakness of the TCE approach. Limited attention is given to exploring other potential governance structures, repeated transactions, the dynamic evolution of governance and transactions, and the key roles of trust and equity in inter-organisational relationships (Ring and van de Ven, 1992). As already stated, the TCE approach argues that hierarchical governance is most appropriate for these kinds of transactions as it reduces the threat of opportunistic behaviour from other parties. The TCE approach fails to recognise that in many industries organisations are involved in complex and collaborative relationships that involve high levels of asset specificity as well as uncertainty and opportunism. In many cases, complex contracts alone cannot guard against the risks associated with uncertainty and opportunism. Relational mechanisms such as trust are regarded as substitutes for complex, explicit contracts and hierarchical govern-ance (Uzzi, 1997; Adler, 2001).

3.3   The resource-based view of the firm

An alternative theory to understanding the boundary of the firm is the resource-based view (RBV). Much of the business strategy literature has been dominated by the RBV that emphasises resources internal to the firm as the principal driver of firm profitability and strategic advantage (Wernerfelt, 1984; Prahalad and Hamel, 1990; Barney,1991; Conner, 1991). The RBV evolved from the work of Edith Penrose (1959) in her book, *The Theory of the Growth of the Firm*, first published in 1959. Her work focused on how firms made decisions on what to produce at what price, and how and why a firm moves from one product and market to another. She identified a number of concepts that are central to the resource-based view of the firm including the accumulation of human and physical resources to exploit services in different products and markets, the ability of managers to exploit these resources and the importance of knowledge to a firm. Resource-based theorists view the firm as a unique bundle of assets and resources that if employed in distinctive ways can create competitive advantage (Barney, 1991; Peteraf, 1993). The distinctive ways in which firms manage these assets and resources can result in superior performance and act as a durable source of competitive advantage. Furthermore, rents derived from services of durable resources that are relatively important to customers and are simultaneously superior, imperfectly imitable, imperfectly substitutable, will not be appropriated if they are non-tradable or traded in imperfect factor markets (Dierickx and Cool, 1989; Barney, 1991; Peteraf, 1993). Teece *et al.* (1997) have proposed an extension to the resource-based view which they term the 'dynamic capabilities' approach. They identify the dimensions of firm-specific capabilities that can be a source of advantage, and explain how combinations of competences and resources can be developed, deployed, and protected. This approach empha-sises how exploiting existing internal and external firm-specific competences can address environmental change. Particular emphasis is given to the development of management capabilities, and difficult-to-imitate combinations of organizational, functional and technological skills.

Barney (2002) has shown that it is possible to relate the resource-based view to analysing the strengths and weaknesses of an organisation. According to Barney (2002), a resource with the potential to create competitive advantage must meet the following four criteria.

3.3.1   Value

This is concerned with determining whether the organisation's resources allow the organisation to react to opportunities or threats in the external environment. Resources and capabilities are considered as valuable if they allow the organisation to exploit both opportunities and counter threats. Therefore, these resources should enable the organisation to meet the factors critical to success in their business environment. Barney (2002) argues that an understanding of value links an analysis of internal capability – i.e. strengths and weaknesses – with an external analysis of opportunities and threats.

3.3.2   Rarity

This involves determining how many competitors possess a valuable resource. Clearly, if a number of competitors possess a resource then it is unlikely to be a source of competitive advantage. Resources with these characteristics are often referred to as threshold resources necessary for survival. A valuable resource that is unique amongst both current and potential competitors is going to be a source of competitive advantage. However, over a period of time valuable and rare resources can become threshold resources as competitors copy or improve upon them.

3.3.3   Imitability

Valuable and rare resources can be a source of competitive advantage. However, it is also important to consider the ease with which competitors can copy a valuable and rare resource possessed by an organisation. In effect, this analysis is concerned with determining the sustainability of the competitive advantage in the resource. Valuable and rare resources can be a source of sustainable competitive advantage if organisations are faced with a cost disadvantage in duplicating the valuable resources of a successful organisation. Resources that exhibit these characteristics are described as *imperfectly imitable* (Lippman and Rumelt, 1982; Barney,1986). Dierickx and Cool (1989) have identified five main factors that can impede imitation of valuable and rare resources.

- *Time-compression diseconomies* – there are additional costs associated with accumulating the required resources under time pressure. For example, an MBA student may not accumulate the same level of knowledge in a 1-year programme as in a 2-year programme even if all inputs other than time are doubled.
- *Asset mass efficiencies* – in some cases, resources can be more costly to accumulate when the organisation's existing stock of the asset is small relative to that of competitors. For example, organisations with a strong research and development capability are often in a stronger position to achieve further technological advances and add to their existing resource base of knowledge than organisations that have limited levels of knowledge.
- *Asset interconnectedness* – the lack of complementary resources can prevent an organisation from accumulating a resource. For example, the capability of an organisation in research and development may be adversely affected by a lack of marketing expertise in generating ideas for new products from both current and potential customers.
- *Asset erosion* – all resource bases can decay over time. For example, research and development capability can decay through technological obsolescence. Clearly, the level of decay in a resource will have to be considered in relation to that of competitors and potential new entrants.
- *Causal ambiguity* – in some cases, it is not possible to explicitly determine the factors or processes that are required to accumulate a required resource. For example, it may be possible to identify the resources of a successful organisation. However, it may be extremely difficult to understand the relationship between the resources and how they create competitive advantage.

3.3.4   Organisation

Barney (2002) argues that a firm must be organised to exploit its resources and capabilities. Organisation includes a number of elements including the reporting structure, management control systems and compensation policies. Amit and Schoemaker (1993) describe these as complementary resources and capabilities because of their limited ability to create competitive advantage on their own. However, when combined with other resources an organisation can achieve competitive advantage. It is important to emphasise that even though a firm may possess a range of valuable, rare and costly to imitate resources ineffective organisation will prevent the full exploitation of these resources.

Some of the proponents of the resource-based view have argued that it is more appropriate to explaining the existence of the firm than the transaction cost economics approach. For example, in Conner's critique (1991) of the transaction cost economics approach, she explained that transaction cost economics emphasised the existence of firms as a way of minimising the opportunistic potential that arise when asset-specific investments are made. Therefore, she argued that transaction cost economics viewed the firm as an avoider of a negative opportunism, while the resource-based theory viewed the firm as a bundle of valuable strategic resources that can be a source of competitive advantage. Reve's (1990) perspective also proposed a resource-based critique of transaction cost economics. Reve (1990) defined strategy as 'the match between a firm's unique resources and its relationship to an ever-changing environment to attain its best performance.' Building on Williamson's work Reve has attempted to define the contractual theory of the firm. Cox (1996) argues this was a major theoretical breakthrough as it took Williamson's view of transaction costs into the realms of 'core competence' and 'fit for purpose' asset specificity. In Reve's approach firms have unique resources (core competences and skills) that they must use responsively and with adaptability to meet the challenges of an ever-changing environment. Reve postulates that assets of high specificity, which are necessary to attain the firm's strategic goals, represent the strategic core of the firm. The strategic core is the *raison d'etre* of the firm, defining its economic rationale within an industry. Using transaction cost analysis Reve (1990) argues that core skills can be of four types:

- *Site specificity* – locational advantage, for example, obtained through locating close to raw material sources;
- *Physical asset specificity* – technology advantages, for example, developed through investments in equipment or technology;
- *Human asset specificity* – know-how advantages, for example, developed over time between engineers and designers in a process; and
- *Dedicated assets* – specialised investments, for example, the development of an information system specifically for the firm with no alternative uses.

A firm must defend these skills if it is to sustain its competitive advantage. Complementary skills can be dealt with through strategic alliances or co-operative relationships if they are of medium asset specificity whilst skills of low asset specificity are left to the market. Reve (1990) argues that the strategic core must be redefined as market and competitive forces continuously change. In a changing business environment a strategic core, which secured competitive advantage in the previous year, may be of no advantage in the current year.

The core competence approach and its relationship with outsourcing have evolved from the resource-based view of the firm. Core competence is important to the study of outsourcing, as it has proposed the internal organisation of the firm as the potential for competitive advantage. Obtaining, creating and developing certain capabilities is central to the core competence approach and has important implications for which activities should be kept within the firm and which should be external to the firm. The ideas of core competence and its relationship to outsourcing have been influenced by the work of Prahalad and Hamel (1990).

They contend that core competencies are not physical assets. Physical assets, no matter how innovative they may seem in the present, can be very easily replicated or become obsolete. Instead, Prahalad and Hamel argue that the real sources of competitive advantage are to be found in management's ability to consolidate corporate-wide technologies and production skills into competencies that empower individual businesses to adapt rapidly to changing business opportunities. They argue that core competencies are 'the collective learning in the organisation, especially how to co-ordinate diverse production skills and integrate multiple streams of technologies'. Competencies are the skills, knowledge and technologies that an organisation possesses on which its success depends. To qualify as a core competence Hamel and Prahalad (1994) argue that it must meet the following three tests.

- *Customer value* – a core competence must enable an organisation to provide a fundamental customer benefit and make a contribution to customer perceived value.
- *Competitor differentiation* – a core competence must be competitively unique and substantially superior to other competitors.
- *Extendability* – as well as meeting the tests of customer value and competitor differentiation, the core competence must be a source of creating an array of products and services in the future. In this way, it should act as a gateway to tomorrow's markets.

Although an organisation will need to reach a threshold level of competence in all activities that it undertakes, it is likely that only some of these activities are core competencies. These core competencies underpin the ability of the organisation to outperform the competition and therefore must be defended and nurtured. In effect, a core competence concerns those resources that are fundamental to a company's strategic position. Instead of developing a strategy based on thinking only of dominating markets, it is more beneficial to think in terms of core competencies, which will segment the organisation in a totally different way. Quinn and Hilmer (1994) argue that core competencies are the following.

- *Skills or knowledge sets, not products or functions* – competencies tend to be skills that cut across business functions. For example, new product development or customer service depend upon cross-functional knowledge rather than the ownership of assets.
- *Flexible, long-term platforms that are capable of adaptation or evolution* – there is considerable value for an organisation in building a myriad of skills that customers will value. For example, Canon's core competencies in optics, imaging and microprocessor controls have allowed it to be a significant player in markets as diverse as photocopiers, laser printers, cameras and image scanners.
- *Limited in number* – it is feasible for an organisation to have only two or three due to the significant investment of time and resources in building core competencies.

- *Unique source of leverage in the value chain* – core competencies should provide an organisation with a significant advantage over their direct competitors in a particular part of the value chain.
- *Areas where the organisation can dominate* – an organisation will excel if it can perform some of the activities – which are important to customers – more effectively than its competitors.
- *Elements important to customers in the long run* – the organisation must specialise and develop a skill or knowledge set that will enable it to serve the needs of customers. By analysing customer requirements, an organisation can specialise and deliver an activity at a lower cost or more effectively to the customer.
- *Embedded in the systems of the organisation* – sustainable competencies will be embedded in functional know-how, processes and technologies that the organisation has developed and nurtured.

The core competence concept has had a significant impact upon many of the outsourcing approaches proposed in the literature (Bettis *et al.*, 1992; Welch and Nayak, 1992; Venkatesan, 1992; Quinn and Hilmer, 1994; Kelley, 1995; Rothery and Robertson, 1995). Much of the literature uses the core competence approach as a starting point for the outsourcing process. For example, Quinn and Hilmer's (1994) analysis of the outsourcing process starts with the core competence approach. Quinn and Hilmer (1994) argue that the firm should concentrate its resources on a set of 'core competences' where it can achieve definable preeminence and provide unique value for customers and strategically outsource other activities for which it has neither a critical strategic need nor special capabilities. Similarly, Venkatesan (1992) argues that the organisation should decide what sub-systems will be indispensable to its competitive position over subsequent product generations. This choice will vary from company to company and ultimately drive product differentiation. In summary, the approaches of these authors consider core, in the context of outsourcing, to have the following characteristics.

- The activity is crucial in the eyes of the customer i.e. one of the key reasons why the customer purchases the product or service.
- The activity may be a source of competitive differentiation in the marketplace.
- The organisation is more competent at performing the activity than suppliers or competitors.

However, there are inconsistencies with these approaches in their interpretation of the core competence concept. For example, some of these approaches can lead to a misinterpretation of the term 'non-core'. Activities that contribute little to competitive differentiation or for which the organisation has no crucial strategic need can be defined as non-core (Venkatesan, 1992; Quinn and Hilmer, 1994). Therefore, such activities should be outsourced and the organisation should focus on activities that can create competitive advantage. However, this analysis implies that all activities procured from the supply market are non-core and thus of no strategic significance. For example, once a company has defined its core competences all other activities are described as non-core and can be sourced externally from more competent suppliers. Therefore, supplier-provided activities are considered to be less important. Accepting this reasoning would imply that all activities or items sourced from the supply market are of no strategic significance. However, it is possible for outsourced activities to be of strategic significance to the organisation if they are still important in the eyes of the customer. This becomes evident if this analysis is used to define a non-core activity. There are a number of conditions when an activity may be termed as non-core (Venkatesan, 1992; Quinn and Hilmer, 1994).

- External suppliers become more competent at performing the activity than the customer organisation.
- A competitor of the customer organisation becomes more competent at performing the activity.

Taking each of these conditions in turn it can be seen that the management of the outsourcing process and supplier relationship is of strategic importance.

- *More competent supplier(s)* – if suppliers become more competent at performing the activity then the organisation may decide to outsource the activity to the more competent supplier. If the organisation selects this alternative it does not automatically follow that the activity becomes non-core or of no strategic significance. It is quite possible that such an activity is still regarded as crucial in the eyes of the customer. In fact, even if the activity is important in the eyes of the customer and there are risks with sourcing externally, it may be more prudent for the company to source the activity externally particularly if there is a rapidly developing supply market. For example, many companies have outsourced their customer service functions to specialists that can create a greater depth of knowledge and be more efficient than many integrated companies.
- *More competent competitor(s)* – if a competitor becomes more competent at performing the activity, this has significant implications for the organisation. The organisation may decide to invest the necessary resources to bridge the disparity between the organisation and the more competent external providers of the activity. This option may be desirable in a case where the technologies involved in the activity are in the embryonic stage and therefore may provide considerable scope for future growth. However, if the organisation's capabilities lag considerably behind the capabilities of external providers, then it may be difficult to justify a substantial investment of resources in order to match or advance upon external capabilities. In this case, the organisation may have no other choice than to outsource the activity. Again, if the organisation decides to outsource such an activity it does not automatically become non-core for the organisation; i.e. of no strategic significance.

Therefore, there is a tendency to associate the term 'non-core' with outsourced activities that have a limited impact on competitive advantage. Under this logic, the term non-core when associated with outsourcing implies that the activity is no longer important in the eyes of the customer and can be provided by a much more capable supplier. However, this analysis undermines the use of the term non-core and makes it an extremely dangerous concept when adopting an outsourcing strategy. It is crucial to clarify why an activity has been outsourced; namely due to more capable suppliers/competitors or no longer a source of competitive differentiation. Each of these instances has different implications for the management of the outsourcing process. For example, activities important in the eyes of the customer and sourced externally in the supply market can still be of strategic importance.

3.4   The industry view

The dominant theory in the field of business strategy during the 1980s was the industry analysis frameworks proposed by Porter (1980). The industry view is most strongly associated with the structure–conduct–performance mode (Mason, 1949; Bain, 1959) that emphasises the actions a firm can take to defend itself against a number of competitive forces. Mason's (1949) analysis of the policies of a number of firms found that differences in market structure influenced the policy decisions taken by firms. Through an investigation of a number of US industries, Bain (1959) challenged traditional economic price theory by introducing the threat of new entrants into a market. By defining entry barriers Bain (1959) introduced much of the language that is used in the strategy fraternity today. The industry analysis approach views business strategy formulation as relating a firm to its environment with the key aspect of the firm's environment being the industry in which it competes. Therefore, the profit potential of a firm is primarily a function of its membership of an industry that exhibits favourable structural characteristics. Porter (1980) has identified five structural characteristics including entry barriers, threat of substitution, bargaining power of buyers, bargaining power of suppliers, and rivalry among competitors in the industry. This approach can assist a firm in finding a position in an industry from which it can best defend itself against these competitive forces or influence them in a way that positively impacts upon its competitive position.

The industry view (IV) has attracted significant attention as a key determinant of firm profitability. A strong position within an industry relative to competitors indicates that the firm has a sustainable competitive advantage. Porter (1985) has identified three strategies for achieving competitive advantage – differentiation, cost leadership and focus – known as the generic strategies. Differentiation

Table 3.1 *The generic drivers (Porter, 1985)*

Differentiation drivers	Cost drivers
• *Policy choices* – related to what activities to perform and how to perform them.	• *Economies of scale.*
• *Linkages*– with customer, suppliers, channels in the value chain.	• *Learning and spillovers* – for example, learning can be exploited in a number of industry contexts.
• *Timing* – e.g. first to market with a product.	• *Patterns of capacity utilisation.*
• *Location.*	• *Linkages.*
• *Inter-relationships* – e.g. cross-functional collaboration.	• *Inter-relationships.*
• *Learning.*	• *Integration.*
• *Integration* – scope and organisation of activities.	• *Timing.*
• *Scale.*	• *Policy choices.*
• *Institutional factors* – government legislation, influence of unions etc.	• *Location*
	• *Institutional factors.*

involves offering customers something unique for which they are willing to pay a premium price. Cost leadership involves being the lowest cost producer in the industry through exploiting economies of scale or other sources cost advantage. A focus strategy involves focusing on a narrow market segment. The firm can either pursue a differentiation or cost leadership strategy in the chosen segment. Porter (1985) has identified a number of generic drivers that can act as a source of either a differentiation or cost leadership strategy. These drivers include scale economies, capacity utilisation, linkages, inter-relationships, vertical integration, location, timing, learning, policy decisions, and government regulations. These are outlined in Table 3.1. Porter (1985) argues that the relative performance between competitors depends upon how each firm exploits these drivers of competitive advantage in an industry. The objective for companies is to position themselves in a more advantageous position in relation to the competitive forces compared with their competitors. For example, if innovative product design is a critical source of differentiation, then those companies with superior design capabilities will have an advantage over their less capable rivals. Alternatively, if economies of scale are a critical source of cost leadership in an industry, then companies that are achieving greater scale relative to competitors will obtain higher levels of profits.

The competitive forces approach provides a framework for analysing the impact of the competitive forces at the industry level. Moreover, this analysis can be used to determine the profitability of different industries and industry segments. This approach to strategy formulation has had a major impact at multi-divisional companies such as General Electric and Shell. Much of strategy formulation at the corporate level of these organisations involves analysing a portfolio of businesses – often referred to as strategic business units (SBUs) – and allocating financial capital as appropriate. For example, in the case of an industry where there is likely to be a flood of new entrants which is likely to reduce the profit potential of the segment, then the organisation may decide to divest the SBU concerned. In effect, decisions in relation to business development are made on the basis of the relative attractiveness of each industry in which the organisation competes.

There has been much debate amongst strategy theorists on the merits of the industry and resource-based views. For example, in discussing the characteristics of the industry view, Teece *et al.* (1997) argues that the strategy process primarily involves choosing rationally amongst a well-defined set of investment options. Resources not owned by the organisation can be readily acquired. Resource-based theorists argue that the resource-based view is a whole new paradigm in understanding competitive advantage in business strategy. The resource-based theorists view the firm as a bundle of resources in the form of both tangible and intangible assets which the firm can exploit. Organisations should focus on developing their resource base and capabilities in order to achieve sustainable competitive advantage. The resource-based perspective views firms with superior systems and structures as being more profitable not because they engage in strategic investments that may deter entry and raise prices, but because they have significantly lower costs, or offer higher quality or superior products. Therefore, this approach focuses on the returns that the firm generates from owning a rare firm-specific resource rather than the profits obtained from selecting an industry.

However, proponents of the industry view such as Porter (1991) argue that the resource-based view cannot be an alternative theory of strategy arguing against the notion that resources are the key to competitive advantage. Much of the debate on the resource-based and industry views has tended to focus on each theorist using data to support their own theoretical standpoint or disprove the other's. For example, some of the resource-based theorists cite the findings of a study undertaken by Richard Rumelt (1991) as support for the resource-based view (Baden-Fuller and Stopford, 1992; Kay, 1995). Rumelt analysed a number of US manufacturing companies between 1974 and 1977 to determine the level of variance in profits that could be explained by factors such as industry participation and corporate ownership. The results of this study are shown in Table 3.2. The key finding of this study for the resource-based theorists was that industry selection had a much less significant impact upon profitability than business unit specific effects. Kay (1995) argues that value is added at the level at which competitive advantage is created – the individual business unit. However, research carried out by McGahan and Porter (1997) challenge some of these findings. Their study was based on data that spanned a longer time period – 1982 to 1994 – and covered a range of industry sectors. They found that industry choice accounted for a greater variation – 19% compared with 8% – in profitability than Rumelt's earlier study had suggested. In the case of service sectors analysed, it was found that industry choice was the most significant influence on profitability accounting for 40–65% of the variance.

Table 3.2 *Causes of variations in profits across business units (Rumelt, 1991)*

Factor	% Effect
Corporate parent ownership	0.8
Choice of industry	8.3
Cyclical effects	7.8
Business unit specific effects	46.4
Unexplained factors	36.7

Despite the increasing dominance of the resource-based view in strategy think-ing in the 1990s, each theory has been addressing the same critical issue – gaining and sustaining competitive advantage – but in different ways. For example, consider a firm that has gained a competitive advantage in a particular industry based on economies of scale. Porter regards this as a valuable strategic resource because it is a barrier to entry into the industry. Alternatively, the resource-based theorists regard this as a valuable resource because a potential new entrant will find it difficult to imitate. However, some argue that the resource-based and industry views can complement one another (Amit and Schoemaker, 1993; Verdin and Williamson, 1994). For example, Amit and Schoemaker (1993) argue that the resource-based view complements the industry view. The latter focuses on product markets viewing the sources of profitability to be the char-acteristics of the industry as well as the firm's position within the industry. The resource-based view holds that the type, magnitude, and nature of firm's resources and capabilities are importance determinants of its profitability. Much of Porter's work also recognises and exhibits similar characteristics to that of the resource-based theory. For example, Porter (1980) has outlined a methodology for analys-ing capabilities in the context of competitor reaction. Porter (1991) has also recognised that distinctive competencies are central to an understanding of strat-egy. Indeed, Porter's (1985) value chain has been recognised as one of the most useful techniques for the analysis of internal resources. One of the most prominent resource-based theorists, Jay Barney, has integrated Porter's value chain into resource analysis – termed the VRIO framework. Barney (2002) argues that activities and how they are linked to one another can be thought of as a firm's resources and capabilities.

3.5   The relational view

A growing body of literature now exists in the area of inter-organisational rela-tionships (Casson, 1998; Dyer and Singh, 1998; Poppo and Zenger, 1998).

Proponents of this literature – sometimes referred to as the relational view (RV) – propose that it is a means of understanding how firms can gain and sustain competitive advantage. For example, Dyer and Singh (1998) argue that it is possible for organisations to combine resources in unique ways across organisational boundaries to obtain an advantage over their competitors. The relational view has evolved from the limitations of TCE in relation to potential governance structures. For example, Ring and van de Ven (1992) have argued that the tendency to focus on markets and hierarchies in the choice of governance in the TCE approach has left a significant void in understanding the potential alternatives. The relational view begins with an understanding of the firm. However, the relational view argues that the firm can develop valuable resources by carefully managing relationships with external entities including suppliers, customers, government agencies and universities. Therefore, a firm can gain and sustain competitive advantage by accessing its key resources in a way that spans the boundaries of the firm. Competitive advantage can be embedded in a set of relationships across the boundaries of firms, rather than residing inside an individual firm. As a consequence, strategy scholars have been analysing the importance of governance mechanisms such as trust and the importance of resources and capabilities of suppliers and customers (Lorenzoni and Lipparini, 1999; Gulati, 1999; Kaufman *et al.*, 2000).

Research has suggested that there is the potential for productivity improvements in the value chain when organisations are willing to make relation-specific investments and combine resources in unique ways (Dyer, 1996). Organisations that make relation-specific investments and are able to combine resources in unique ways to generate relational rents and gain competitive advantage over organisations that are unable to do this. In proposing the relational view, Dyer and Singh (1998) define relational rents as profits generated jointly in the relationship that cannot be generated by either firm alone and can only be created through the joint contributions of the specific alliance partners. Dyer and Singh (1998) have identified four sources of relational rents.

- *Inter-firm specific assets* – this involves creating specialised assets in conjunction with the assets of an alliance firm. This is similar to the concept of asset specificity. For example, relational rents can be generated through long-term relationships in which supplier designers learn the systems and procedures that are specific to the buyer.
- *Inter-firm knowledge-sharing routines* – inter-firm processes can be designed to facilitate knowledge exchanges between the alliance firms. For example, Toyota transfers knowledge directly to suppliers, through its 'operations management consultancy division' consultants, who may reside at the supplier organisation for days, weeks, or months to ensure that the transfer takes place (Nishiguchi, 1994).

- *Complementary resource endowments* – these are defined as distinctive resources of alliance firms that collectively generate rents greater than the sum of those obtained from the individual endowments of each partner (Dyer and Singh, 1998). For example, a company with a strongly branded product portfolio may establish an alliance with a company which possesses a global distribution network. In this case, both companies bring distinctive competences to the alliance, which, when combined with the resources of the partner create combined resource endowments that are more rare, valuable, and difficult to imitate than they had been before they were combined.
- *Effective governance* – governance is central to the creation of relational rents because of its impact upon both transaction costs and value creation in the relationship. Dyer and Singh (1998) distinguish between two types of governance in the context of alliances either third-party legal enforcement of agreements or self-enforcing agreements in which disputes are resolved through relational mechanisms. Effective governance can generate relational rents by either reducing transaction costs or developing incentives to create value in the relationship through knowledge sharing or combining complementary resources.

The relational view is closely related to some of the concepts associated with knowledge-based theory (Lorenzoni and Lipparini, 1999). For example, the buyer and supplier can create tacit knowledge through greater collaboration in new product development and the application of information technologies to facilitate greater process integration. The creation of knowledge in the relationship can also be a source of innovation. In many industries, most innovation occurs at the inter-organisational level with customers making more than 50% of innovations on new products. For example, Powell *et al.* (1996) found that the principal source of innovation in the biotechnology industry was in the linkages between collaborating organisations and not the single organisation. Patents were normally filed by more than one organisation from a number of areas including biotechnology, pharmaceuticals and academia. Organisations involved in this type of collaboration can develop a relationship that delivers benefits unavailable to competitors through the deployment of mechanisms to facilitate knowledge sharing. Grant (1996) defines these mechanisms as inter-firm knowledge-sharing routines that involve a regular pattern of inter-firm interactions that allow the transfer, recombination, or creation of specialised knowledge.

The importance of inter-organisational relationships has led to a growing body of literature in the area of supply chain relationships (Harland, 1996; Ellram and Edis, 1996; Bensaou, 1999). Increasingly, companies have been pursuing more intensive and interactive relationships with their suppliers, collaborating in areas such as new product development, supplier development, and information sharing on a range of issues (McIvor *et al.*, 1997). Furthermore, the trend towards increased outsourcing has led companies to become more dependent on their supplier network. Much of the force behind Western companies pursuing more collaborative buyer-supplier relationships has come from the observation of the success of Japanese management practices. Japanese companies have tended to outsource more activities and required suppliers to invest in relationship-specific physical and human capital (Nishiguchi, 1994). Due to the success attributed to Japanese-style partnerships and the emergence of Japanese transplants such as Honda and Nissan in the US and UK, Western companies have become interested in adopting many of the features of partnership sourcing (Macbeth, 1994; McIvor *et al.*, 1997; Pickernell, 1997). For example, collaboration has been a strong feature of Nissan's relationships with its suppliers and a non-confrontational approach has been prominent from the beginning in Sunderland. Much of the writing in supply chain relationships in academia has evolved from analyses of the experiences of the automotive industry and has led to the development of concepts such as partnership sourcing (Macbeth, 1994), network sourcing (Hines, 1994), and lean supply (Lamming, 1993).

3.6   Integrating the theoretical influences into outsourcing evaluation and management

The previous sections have outlined both the characteristics and limitations of the TCE approach, the RBV, the IV and RV. This section argues that there are a number of interdependencies with each of these theoretical perspectives that can assist in outsourcing evaluation and management.

3.6.1   The RBV and TCE

The TCE approach and the RBV have already been discussed in relation to their potential for understanding the boundary of the organisation. Essentially, the TCE approach is addressing why firms exist and the RBV considers why firms differ. Therefore, the TCE approach focuses primarily on the role of efficient governance – through transaction analysis – in explaining firms as institutions for organising economic activity, whilst the RBV focuses on the search for competitive advantage – through resource analysis. The TCE approach is focusing primarily on governance skills whilst the RBV focuses primarily on production skills. However, it is important to consider the inter-dependencies of the TCE approach and the RBV to understanding outsourcing evaluation and in particular, the relationship between production and governance skills. The RBV assists in the analysis of production skills. For example, production skills refer to the routines, processes and knowledge required to build valuable strategic resources. In relation to understanding governance, the TCE approach proposes that the most appropriate mode of governance is based on the costs of governance and the threat of opportunism. Therefore, integrating these two perspectives into the context of outsourcing, an organisation can access the required production skills either – internally or externally from a supplier – through employing the most appropriate governance mode. This analysis is particularly useful in the situation where there is a clear distinction between markets and hierarchies as suitable governance modes.

3.6.2   The RBV and the RV

To further strengthen this analysis it is important to emphasise that in some contexts inter-organisational collaboration can be employed to access and develop complementary resources that are required for business success. As illustrated earlier, in some contexts such as rapidly changing industries in which organisations have to access certain capabilities in order to compete, the TCE approach does not fully explain all the potential governance alternatives. However, governance skills, both internally and externally across organisational boundaries, can lead to superior performance and competitive advantage. The buyer and supplier are jointly involved in both production and governance. This contrasts with the TCE approach that argues that collaborative relations are employed solely to reduce the cost of governing the resource. Therefore, the sole focus on cost of governance limits the potential benefits and value that can be created from employing collaborative relationships. Inter-organisational relationships characterised by high levels of transaction-specific investments are most likely to create valuable, rare, and costly-to-imitate resources and capabilities for a firm. Madhok (2002) argues that the key issue is the net value of the transaction rather than the transactions costs. Whilst there are costs associated with linking the two sets of resources together, the value of the transaction is derived from the economic surplus through integrating together production and exchange relations which could more than offset the associated costs. Possessing the appropriate governance skills can allow an organisation to access a wider range of resources outside the firm. The ability to employ the range of available governance alternatives more astutely than competitors can be a source of competitive advantage. In fact, governance can become one of the most critical resources the firm possesses.

3.6.3   The RBV and IV

The resource-based view of the firm is based on the building of rare, valuable, non-substitutable and difficult-to imitate resources to act as a source of competitive advantage (Barney, 1995). The resource-based view is important to the study of outsourcing as it contends that the reason an activity is performed internally within the firm is because it is a source of competitive advantage. Therefore, in the context of outsourcing, organisations should focus on creating these critical strategic resources in order to achieve competitive advantage. To further reinforce this analysis in identifying critical resources, certain aspects of the industry view can be integrated into this analysis. Although some proponents of the resource-based and industry views argue that the two theories are exclusive, integrating certain aspects of each of the theories can assist in the outsourcing evaluation process. In particular, it is possible to link Porter's (1985) generic drivers that can act as a source of competitive advantage to the resources that are required to gain and sustain competitive advantage.

A key part of outsourcing evaluation involves determining the critical activities of the business: i.e. those activities that can be a source of competitive advantage. For example, if an organisation possesses a distinctive capability in a critical activity then it should be performed internally. This evaluation can either begin with an internal analysis of organisational capabilities in order to determine the unique resources (the resource-based approach) or external analysis of the competitive environment to determine the drivers of competitive advantage (the industry view). However, both of these approaches are searching for the same outcome. Whether the analysis focuses on the internal or external dimension, the organisational activities that underpin the drivers of competitive advantage will have to be determined. Clearly, activities that underpin a generic driver and in which the organisation has a strong position will be deemed critical and have to be controlled and developed internally. Alternatively, the organisation may consider outsourcing an activity that is not critical to competitive advantage. Furthermore, the analysis may reveal a potential driver of competitive advantage for which the organisation does not possess the underpinning activity. For example, an organisation with a limited Internet presence may find that the Internet has become a key source of advantage for adding value to its product and service portfolio both currently and in the future. In this case, the organisation will have to consider acquiring the capability or employing relational mechanisms with a supplier that possesses the capability. Also, it is stressing that a number of the generic drivers of competitive advantage (including 'linkages' and 'interrelationships') identified by Porter (1985) involve the development of inter-organisational relationships.

REFERENCES

Adler, P. (2001). Market, hierarchy, and trust: the knowledge economy and future of capitalism. *Organisation Science*, **12**, No. 2, 214–34.

Amit, R. and Schoemaker, P. J. H. (1993). Strategic assets and organisational Rent. *Strategic Management Journal*, **14**, No. 1, 33–45.

Baden-Fuller, C. and Stopford, J. M. (1992). *Rejuvenating the Mature Business: The Competitive Challenge*. Cambridge, MA: Harvard Business School Press.

Bain, J. S. (1959). *Industrial Organisation*. New York: Wiley.

Barney, J. B. (1986). Organisational culture: can it be a source of sustained competitive advantage? *Academy of Management Review*, **11**, 656–65.

Barney, J. B. (1991). Firm resources and sustained competitive advantage. *Journal of Management*, **17**, No. 1, 99–120.

Barney, J. B. (1995). Looking inside the organisation for competitive advantage. *Academy of Management Executive*, **9**, 49–61.

Barney, J. B. (1999). How a firm's capabilities affect boundary decisions. *Sloan Management Review*, **40**, 3, 137–45.

Barney, J. A. (2002). *Gaining and Sustaining Competitive Advantage*. New Jersey: Prentice Hall, second edition.

Bensaou, M. (1999). Portfolios of buyer–supplier relationships. *Sloan Management Review*, **39**, No. 4, 35–44.

Bettis, R. A., Bradley, S. P., & Hamel, G. (1992). Outsourcing and industrial decline. *Academy of Management Executive*, **6**, No. 1,7–22.

Casson, M. (1998). *Information and Organisation: A New Perspective on the Theory of the Firm*. New York: Oxford University Press.

Conner, K. R. (1991). A historical comparison of resource-based theory and five schools of thought within industrial organization economics: do we have a new theory of the firm? *Journal of Management*, **17**, No. 1, 121–54.

Cox, A. (1996). Relational competence analysis and strategic procurement management: towards an entrepreneurial and contractual theory of the firm. *European Journal of Purchasing & Supply Management*, **2**, No. 1, 57–70.

Dierickx, I. and Cool, K. (1989). Asset stock accumulation and sustainability of competitive advantage. *Management Science*, **35**, No. 2, 1504–11.

Doz, I. and Prahalad, C. K. (1991). Managing DMNCs: a search for a new paradigm. *Strategic Management Journal*, **12**, Special issue, 145–64.

Dyer, J. H. (1996). Specialised supplier networks as a source of competitive advantage: evidence from the auto industry. *Strategic Management Journal*, **17**, 271–92.

Dyer, J. H. and Singh, H. (1998). The relational view: cooperative strategy and sources of inter-organisational competitive advantage. *Academy of Management Review*, No. **23**, 660–79.

Ellram, L. M. and Edis, O. R. V. (1996). A case study of successful partnering implementation. *International Journal of Purchasing and Materials Management*, **32**, No. 3, 20–8.

Grant, R. (1996). Prospering in dynamically competitive environments: organisation capability as knowledge integration. *Organisation Science*, **7**, 375–87.

Gulati, R. (1999). Network location and learning: the influence of network resources and capabilities on alliance formation. *Strategic Management Journal*, **20**, No. 5, 397–420.

Hamel, G., and Prahalad, C. K. (1994). *Competing for the Future*. Boston: Harvard Business Press.

Harland, C. M. (1996). Supply chain management: relationships, chains and networks. *British Journal of Management*, **7**, 63–80.

Hines, P. (1994). *Creating World Class Suppliers: Unlocking Mutual Competitive Advantage*. London: Pitman.

Kaufman, A., Wood, C. H. and Theyel, G. (2000). Collaboration and technology linkages: a strategic supplier typology. *Strategic Management Journal*, **21**, 649–63.

Kay, J. (1995). *Foundations of Corporate Success.* Oxford: Oxford University Press.

Kelley, B. (1995). Outsourcing marches on. *Journal of Business Strategy,* **16**, No. 4, 38–43.

Lamming, R. (1993). *Beyond Partnership, Strategies for Innovation and Lean Supply.* Hemel Hempstead, UK: Prentice-Hall.

Lippman, S., and Rumelt, R. P. (1982). Uncertain imitability: an analysis of inter-firm differences in efficiency under competition. *Bell Journal of Economics,* **13**, 418–38.

Lorenzoni, G. and Lipparini, A. (1999). The leveraging of inter-firm relationships as a distinctive organisational capability: a longitudinal study. *Strategic Management Journal,* **20**, No. 4, 317–38.

MacBeth, D. K. (1994). The role of purchasing in a partnering relationship. *The European Journal of purchasing and Supply Management,* **1**, No. 1,19–25.

Madhok, A. (2002). Reassessing the fundamentals and beyond: Ronald Coase, the transaction cost and resource-based theories of the firm and institutional structure of production. *Strategic Management Journal,* **23**, 535–50.

Mason, E. (1949). The current state of the monopoly problem in the US. *Harvard Law Review,* **62**, 1265–85.

McGahan, A. M., and Porter, M. E. (1997). How much does industry matter, really? *Strategic Management Journal,* **18**, 15–30.

McIvor, R. T., Humphreys, P. K. and McAleer, W. E. (1997). The implications of the trend towards partnership sourcing on buyer–supplier relations. *The Journal of General Management,* **23**, No. 1, 53–69.

Nishiguchi, T. (1994). *Strategic Industrial Sourcing.,* New York: Oxford University Press.

Penrose, E. (1959). *The Theory of the Growth of the Firm.* Oxford: Blackwell.

Peteraf, M. A. (1993). The cornerstones of competitive advantage: a resource-based view. *Strategic Management Journal,* **14**, 179–91.

Pickernell, D. (1997). Less pain but what gain: a comparison of the effectiveness and effects of Japanese and non-Japanese car assemblers buyer–supplier relations in the UK automotive industry. *OMEGA,* **25**, No. 4, 377–95.

Poppo, L. and Zenger, T. (1998). Testing alternative theories of the firm: transaction cost, knowledge-based and measurement explanations of make-or-buy decisions in information services. *Strategic Management Journal,* **19**, No. 9, 853–77.

Poppo, L. and Zenger, T. (2002). Do formal contracts and relational governance function as substitutes or complements? *Strategic Management Journal,* **23**, 707–25.

Porter, M. E. (1980). *Competitive Strategy.* New York: Free Press.

Porter, M. E. (1985). *Competitive Advantage: Creating and Sustaining Superior Performance.* New York: Free Press.

Porter, M. E. (1991). Towards a dynamic theory of strategy. *Strategic Management Journal,* **12**, 95–117.

Powell, W., Koput, K and Smith-Doerr, L. (1996). Inter-organisational collaboration and the locus of innovation: networks of learning in biotechnology. *Administrative Science Quarterly,* **41**, 116–45.

Prahalad, C. K. and Hamel, G. (1990). The core competence of the corporation. *Harvard Business Review,* **68**, No. 4, 79–91.

Quinn, J. B. and Hilmer, F. G. (1994). Strategic outsourcing. *Sloan Management Review,* **35**, No. 4, 43–55.

Reve, T. (1990). The firm as a nexus of internal and external contracts. In: *The Firm as a Nexus of Treaties,* Aoki, M. (ed.). London: Sage.

Ring, P. S. and van de Ven, A. H. (1992). Structuring cooperative relationships between organisations. *Strategic Management Journal*, **13**, 483–98.

Rothery, B. and Robertson, I. (1995). *The Truth about Outsourcing*. Aldershot,UK: Gower Publishing.

Rumelt, R. P. (1991). How much does industry matter? *Strategic Management Journal*, **12**, No. 3, 167–86.

Teece, D. J., Pisano, G. and Shuen, A. (1997). Dynamic capabilities and strategic management. *Strategic Management Journal*, **18**, No. 7, 509–34.

Uzzi, B. (1997). Social structure and competition in inter-firm networks: the paradox of embeddedness. *Administrative Science Quarterly*, **42**, 35–67.

Venkatesan, R. (1992). Strategic sourcing: to make or not to make. *Harvard Business Review*, **70**, No. 6, 98–107.

Verdin, P. J. and Williamson, P. J. (1994). Core competences, competitive advantage and market analysis: forging the links. In: Competence-based Competition, Hamel, G. and Heene, A. (eds.). Chichester: Wiley, pp. 77–110.

Welch, J. A. and Nayak, P. R. (1992). Strategic sourcing: a progressive approach to the make or buy decision. *Academy of Management Executive*, **6**, No. 1, 23–30.

Wernerfelt, B. (1984). A resource-based view of the firm. *Strategic Management Journal*, **5**, No. 2, 171–80.

Williamson, O. E. (1985). *The Economic Institutions of Capitalism: Firms, Markets and Relational Contracting*. New York: Free Press.

Williamson, O. E. (1975). *Markets and Hierarchies*. New York: Free Press.

4 The outsourcing process: a framework for evaluation and management

4.1 Introduction

This chapter provides an overview of the stages involved in evaluating and managing the outsourcing process. Background information outlining the stages in the development of the outsourcing framework is presented. The initial analysis in the development of the framework revealed a number of problems encountered by organisations in their efforts to both evaluate the suitability of outsourcing for their organisation and the management of the outsourcing process. These problems stemmed from the lack of a structured step-by-step approach to the evaluation and management of outsourcing. In addition, it was found that many organisations were misinterpreting the core competence approach, which has had a major influence on outsourcing both in theory and practice. It is shown how the outsourcing framework is influenced by a number of the theoretical perspectives discussed in the previous chapter including transaction cost economics (TCE), the resource-based view (RBV) of the firm, the industry view (IV) and the relational view (RV). An outline of the issues that should be considered in the key stages of the outsourcing framework is presented. This section also provides an outline of the structure of the book in the following chapters. Finally, the theoretical implications of the outsourcing framework are discussed.

4.2 Practical problems with the outsourcing process

There is evidence to suggest that organisations have not been achieving the desired benefits from outsourcing. For example, three-quarters of respondents to an American Management Association survey reported that the anticipated outcomes of outsourcing had failed to meet their expectations, and more than half reported that they had to bring at least one activity back in-house (Greenberg and Canzoneri, 1997). In fact, it has been argued that outsourcing decisions are made most frequently by default with little consideration for the long run competitiveness of the organisation. Few companies have taken a strategic view of outsourcing decisions, with many companies deciding to buy rather than make for short-term reasons of cost reduction and capacity (McIvor, 2000). In addition, some organisations may find themselves with an initial position that has been inherited from the past. Their position in the supply chain is already established and the extent of vertical and horizontal integration already mapped out. However, this is likely to have occurred due to a series of short-term decisions with no consideration for the long-term strategic direction of the organisation. An outline of three key problems encountered by organisations in their efforts to formulate an effective outsourcing decision is presented below.

4.2.1 No formal outsourcing process

Many organisations have no structured basis for evaluating the outsourcing decision and in particular, do not place the decision in a strategic context. Lonsdale and Cox (1997) have found that many organisations make outsourcing decisions primarily on the basis of reducing headcount and costs. The choice of which parts of the business to outsource is made by ascertaining what will save most on overhead costs, rather than on what makes the most long-term business sense. This can create a situation in which the organisation is outsourcing activities that are a source of competitive advantage. Organisations are failing to consider issues such as the following.

- Should the organisation strive to maintain and build its capability in a particular technology or turn to the best-in-class source?
- Does the necessary capacity exist within the organisation to provide the activity?
- Do the internal capabilities of the organisation in a particular technology lag behind potential suppliers?
- If there is a disparity between purchaser and supplier, how much investment is required internally to match the capabilities of the suppliers?

Outsourcing by a wide variety of organisations has grown over the last few years. The reason for this trend appears to be the disadvantages associated with performing too many activities internally, owing to rapid changes in the market and the lack of flexibility that characterises carrying out too many activities internally (Linder et al., 2002). However, in a manufacturing context it has been found that increased outsourcing of components formerly made internally can result in unexpected cost increases, with many companies failing to integrate the make or buy decision into the overall manufacturing strategy (Probert, 1996). This may lead to a situation in which patches of manufacturing processes are dispersed at random throughout a company's operations and the company becomes more dependent on a much wider range of suppliers.

4.2.2   Insufficient understanding of the costs associated with outsourcing

Many organisations have failed to understand all the major costs associated with outsourcing evaluation and management. For example, often organisations have underestimated the costs of finding and evaluating suppliers, drafting the contract and managing the relationship. Such costs can significantly reduce the financial benefits associated with outsourcing an activity to a supplier. Cost analysis in outsourcing evaluation and management involves attempting to measure all the important costs associated with the two alternatives – perform internally or outsource. The alternative that yields the lowest total cost is chosen. However, in many cases it is not possible to objectively assess all the costs associated with outsourcing. For example, other more qualitative factors, such as the long-term strategic implications for the organisation and workforce reaction to outsourcing may have a greater impact on whether the desired benefits are attained from outsourcing. The problem with evaluating sourcing decisions primarily upon the basis of costs is further exacerbated by the fact that many companies have inadequate costing systems. The results of studies carried out on the cost accounting practices and financial performance systems used by many manufacturers has shown that many of these organisations' accounting systems have not kept pace with industry changes and the technology used in production (Lebas, 1999). For example, direct labour hours is still widely used as the basis for allocating overhead, even while the production process is highly automated. Furthermore, the use of a simplistic allocation of indirect costs on the basis of direct labour cost has been inappropriate for many organisations, particularly those with highly automated processes or complex products with short lifecycles. This situation has led organisations to choose a strategy of de-emphasising and overpricing products that are highly profitable, and expanding commitments to complex, unprofitable lines.

As well as considering the costs of performing the activity internally in relation to potential external suppliers, costs also arise in both searching for suitable suppliers and contracting. These costs are incurred before outsourcing occurs. In searching for suitable suppliers, the sourcing organisation incurs costs in the area of collecting information to identify and evaluate suitable suppliers. Contracting costs refer to the costs associated with negotiating and drawing up a contract with the suitable supplier identified. Additionally, once the sourcing organisation has decided to outsource, there are costs associated with managing the chosen supplier. These costs have three important dimensions (Barthlelemy, 2003), namely:

- monitoring the agreement with the supplier to ensure that contractual obligations are being met;
- bargaining with the supplier and imposing penalties when the supplier fails to fulfil its contractual obligations; and
- negotiating changes to the contract in the case of any unexpected changes in the business environment.

Illustration 4.1

The hidden costs of outsourcing

Jerome Barthelemy analysed the experiences of a number of companies that had been involved in information technology outsourcing. Although many of the companies analysed had obtained benefits in the form of reduced costs and improved performance, other companies who had unsuccessful experiences had failed to understand all the major costs associated with outsourcing. In many cases, these hidden costs eliminated any potential cost savings or performance improvements that these companies were realising through outsourcing. From his analysis Barthelemy identified the following four hidden costs associated with outsourcing.

- *Vendor search and contracting* – many organisations failed to estimate the costs associated with identifying, and evaluating suitable vendors, selecting a suitable vendor, and negotiating a contract.
- *Transitioning to the vendor* – it was found that transitioning the activity being outsourced to the vendor was the most elusive of hidden costs. In fact, many organisations did not realise how much cost was incurred until the transition was complete. For example, it can take months before the vendor is as familiar with the activity as the outsourcing organisation. Furthermore, some organisations found it difficult to determine exactly when the vendor had actually taken over.
- *Managing the effort* – this involves a number of areas including monitoring to determine whether the vendor fulfils their contractual obligations, bargaining with vendors, and negotiating any required contract changes. These costs represent the largest category of hidden costs.
- *Transitioning after outsourcing* – this involves the costs associated with switching vendors or bringing the activity back in-house. Again, this is an extremely difficult cost to quantify. In the case of switching to a new vendor, the costs involved finding the vendor, drafting a new contract and transitioning resources, whilst bringing an activity back in-house signals a failure in managing the outsourcing process effectively.

It must be stressed that many of these costs are extremely difficult if not impossible to measure objectively. However, the benefit of attempting to estimate these potential hidden costs at the outset is that they can challenge the basis for an ill-thought outsourcing process. Barthelemy provides some guidance on how to reduce the potential likelihood of some of these hidden costs arising.

- An organisation should select activities that are relatively straightforward to outsource and not surrounded by too much uncertainty. In fact, the nature of the activity and the experience of the organisation with outsourcing are significant determinants of whether hidden costs are likely to arise.

- Considerable effort should be given to researching and evaluating the capabilities of potential vendors.
- Advice and knowledge should be sought from experts with relevant experience in outsourcing. In many cases, these experts can identify how the pitfalls associated with outsourcing can be avoided. For example, legal advice can be employed to assist in writing the outsourcing contract and negotiating with the vendor. Technical advice can also be sought to develop measures in assessing service levels.
- It is important to draft a clear contract. Poorly drafted contracts will lead to high bargaining and renegotiating costs. In particular, they will make it extremely difficult switch vendors. In the context of information technology outsourcing, Barthelemy argues that contracts should incorporate clauses that specify technology and price evolution and reversibility. Technology and price-flexibility clauses are normally derived from benchmarking clauses; namely comparisons with other competing suppliers. Reversibility clauses can be either material or human. For example, material reversibility allows the outsourcing organisation to buy back the premises or equipment, whereas human reversibility allows the outsourcing organisation to hire employees from the vendor.
- Developing a healthy relationship with the vendor characterised by high levels of trust and joint problem solving can compensate for any gaps in the contract.
- Key people who have in-depth knowledge of the activity should be retained in the outsourcing company. The loss of key people can make moving the activity to another supplier or bringing it back in-house extremely difficult.

Source: Barthelemy, J. (2001). The hidden costs of IT outsourcing. *Sloan Management Review*, **42**, No. 3, 60–9.

4.2.3   Core business approach

The use of the core competence approach has also dominated the approaches to outsourcing by practitioners. However, operationalising the core competence concept has proved difficult as the concepts involve, 'skills/ knowledge sets,' 'root systems,' 'unique sources of leverage,' 'elements important in the long term,' which appear vague and intangible. Due to these difficulties, many companies have unknowingly relinquished their core competencies by cutting internal investment in what they mistakenly thought were 'cost centres' in favour of outside suppliers. In other words, the need to focus on core competencies has been operationalised for cost reduction reasons through the reduction of headcount. Outsourcing may provide a shortcut to a more competitive product, but it typically contributes little to build the people-embodied skills that are needed to sustain future product leadership. There is also evidence to suggest that organisations are misusing the term 'core'. For example, Lonsdale and Cox (1997) have found that some organisations have been defining core activities as 'those things that we do best'. Such an application has clear risks in that it may lead companies to outsourcing activities with which they are experiencing poor performance. Organisations may decide that it is not necessary to allocate resources to tackle activities with which they are experiencing problems. In certain circumstances, where poor performance is a result of a lack of scale economies or knowledge of the process, outsourcing the activity may be appropriate. For example, where a supplier can realise significant scale economies in a process through servicing the requirements of a number of customers, it is very difficult for the sourcing organisation to replicate such a position. However, in circumstances where poor performance is a result of poor management that can be addressed through an internal improvement initiative, then outsourcing can be fraught with risks. For example, these activities may be of significant value to the company currently and in the future and potentially contributing to the organisation's competitive advantage. There are a number of other reasons why strict adherence to the core competence approach to outsourcing can be problematic. The nature of some industries can make the emphasis on the distinction between core and non-core activities as the key motive for outsourcing inappropriate. For example, in rapidly changing industries technological innovations can render an advantage in an activity obsolete almost overnight. In these circumstances, outsourcing is driven more by the nature of the outsourcing contract and the management of the relationship between the sourcing organisation and the supplier. Indeed, as has already been emphasised, the leveraging of supplier capabilities through astute relationship mechanisms can be a source of competitive advantage. Moreover, even when an activity is considered to be 'non-core' to the organisation achieving competitive advantage, there are often implicit and tacit inter-dependencies with activities that are considered to be 'core' (Bryce and Usseem, 1998). For example, when outsourcing an activity an organisation may fail to account for both the formal and informal co-ordinating mechanisms that have allowed the organisation to perform the activity internally in the past. The outsourcing of such an activity can also have a detrimental effect upon the performance of core activities that remain within the organisation. Indeed, the failure to account for the tacit inter-dependencies between organisational activities is a potential risk of outsourcing in general.

4.3   Background to the development of the outsourcing framework

The analysis for the development of the outsourcing framework involved a thorough review of the literature in a number of areas including business strategy, economics, and inter-organisational relationships. The next phase involved talking directly with practitioners to elicit their views on the key steps involved in outsourcing evaluation and management. A series of structured interviews with senior managers in a number of organisations was conducted. A number of interviews were carried out with senior managers from a range of business functions in each organisation in order to obtain a cross-functional perspective on the outsourcing process. These organisations came from a variety of industries including electronics, telecommunications, mechanical engineering, aerospace, chemicals and medical packaging. Issues addressed during the interviews included the following:

- influences on the outsourcing process;
- level of integration with business strategy;
- functions involved in the decision making process;
- level of cost analysis; and
- role of suppliers in the process.

As a result of these interviews and a review of the literature, a framework for the outsourcing evaluation and management process was developed. The framework encompasses a number of variables and seeks to capture the complexities of outsourcing evaluation and management. It identifies the key issues that should be addressed when evaluating the implications of outsourcing for organisations. An explanation is given on how and why these dimensions should be considered in the formulation of the outsourcing decision. The framework seeks to assist both academics and practitioners in understanding the implications of a number of key influences on outsourcing decision-making such as organisational capability, supplier capability, competitor actions, and supply market risk and offers a number of alternatives based upon an analysis of these key influences. This framework contains prescriptive elements because it suggests that the outsourcing process should be carried out in a certain way. A framework is concerned with making recommendations of what to do and what should be done (Hogwood and Gunn, 1984). The outsourcing framework consists of a number of logically sequential steps. There are also explanatory elements within the framework. An explanatory approach describes how things are in the world that is being perceived and constructed. A management consultant might create a report based upon what has been found in an analysis of a set of existing circumstances. For example, a report might document current practice with regard to strategic planning and describe the relationships between the people involved in an organisation. The outsourcing framework here contains elements that describe 'how things are in the world' by drawing upon existing literature and empirical evidence through the structured interviews carried out with senior managers in a range of organisations. It is quite common for frameworks to have both prescriptive and explanatory elements (Saunders, 1997). The adoption of a framework tends to make assumptions about circumstances in which it might be employed. Similarly, explanatory frameworks often contain implicit or explicit normative recommendations.

4.4   A description of the stages in outsourcing evaluation and management

The stages involved in the outsourcing framework are illustrated on the decision tree in Figure 4.1. An overview of the stages involved in evaluating and managing the outsourcing process is now presented.

4.4.1   Stage 1 – Determining the current boundary of the organisation

This stage is concerned with identifying the key activities that are performed in order to create and deliver the range of products and services offered by the organisation. This involves identifying the key activities that the organisation performs internally as well as those performed by external sources. The purpose of this analysis is to provide an outline of the scope of the organisation both upstream into the supply chain and downstream towards end customers of the products and services. This analysis is carried out at the activity level. Porter (1991) defines an organisational activity as the basic unit of competitive advantage. The economics of performing activities will determine the relative cost position of the organisation and not the attributes of the organisation as a whole. Porter (1991) also argues that the organisation can achieve competitive advantage either through performing activities at a lower cost relative to competitors or perform activities in a way that differentiates its product and service offering and for which customers are willing to pay a premium price. There are a number of models for mapping the activities of an organisation. Porter's value chain model represents an organisation as a chain of activities for transforming inputs into outputs that customers value (Porter, 1985). The value shop model proposed by Stabell and Fjeldstadt (1998) can be applied in the context of professional service such as a law firm or architecture practice. Moreover, the value network model is applicable in the context of network-type organisations such as banks and insurance companies. The depth of evaluation can take place at the activity level (such as logistics) or sub-activity level (such as material handling, incoming quality inspection, warehousing etc.) depending upon the particular circumstances of the organisation. Segmentation of an organisation into activities and sub-activities is valuable from the following perspectives.

- It can assist in identifying activities that have a significant influence on whether the organisation's customers buy their products or services. For example, many organisations are investing considerable resource in their customer service operations as it has the potential to act as a source of competitive differentiation.

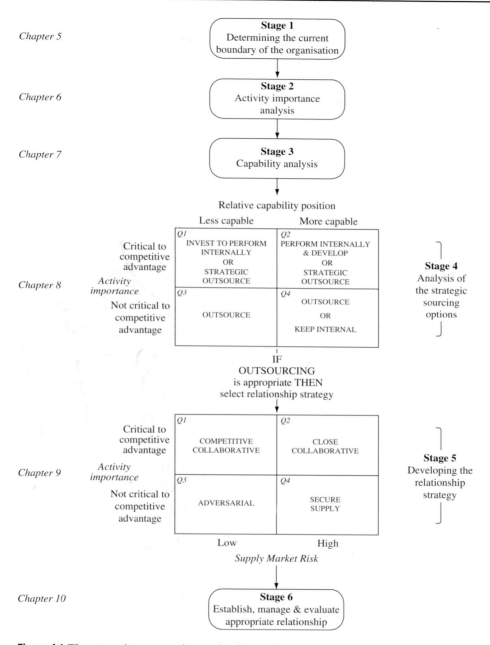

Figure 4.1 The stages in outsourcing evaluation and management

- It can identify activities that represent a significant or growing proportion of cost of producing the product or delivering the service. For example, sub-assembly activities in manufacturing can be the largest single cost associated with the manufacture and delivery of the product to the customer.

When an organisation is viewed from an activity perspective, it is much easier to recognise the value-adding activities that contribute to the organisation's competitive position. For example, in the outsourcing decision for a manufacturing process, it is possible to consider all the associated 'support' activities such as logistics and material handling skills rather than the skills associated with the manufacturing process alone. As well as segmenting the organisation into a number of activities and sub-activities, it is necessary to understand the linkages between activities both inside and outside the organisation. For example, there are important linkages between internal business functions and those of supplier organisations.

4.4.2   Stage 2 – analysis of the importance of organisational activities

This stage in the analysis is concerned with identifying the level of importance of the activities involved in creating and delivering an organisation's range of products and services to customers. A critical activity has a major impact upon the ability of an organisation to achieve competitive advantage. Therefore, the activity is perceived by customers as adding value and offering a unique value proposition. Determining the importance level of activities is central to outsourcing evaluation. For example, a critical activity in which the organisation possesses a distinctive capability will have to receive a considerable level of strategic attention in order to maintain such a position. The first phase in this stage of the analysis is to understand the nature of the competition in the business environment. Porter's Five Forces model can be used to understand the nature of competition. Porter (1980) argues that competition in an industry is rooted in its underlying economics and competitive forces that can extend well beyond the established competitors in a particular industry. For example, customers, suppliers, new entrants, and substitute products or services are all potential competitors with their competitive stance being significantly influenced by the industry under analysis. Once this analysis is carried out, the organisation should undertake an analysis of the current and future determinants of success in the business environment. There are a number of techniques that can be used in this phase of the analysis.

- *The value concept* – understanding the concept of customer value is a useful basis for determining the importance level of organisational activities. Normann and Ramirez (1993) have argued that strategy is 'the art of creating value'. Business strategy provides the intellectual frameworks, conceptual models and governing ideas that enable organisations to identify opportunities for creating value for customers and for delivering that value at a profit. A useful approach for understanding value is the 'perceived use value' approach proposed by Bowman (1998) who defines the value of a product or service as the perceptions that a customer has of the usefulness of the product or service.

- *The critical success factor (CSF) method* – identifying CSFs can be also used to determine the importance level of organisational activities. CSFs are those things that the organisation must perform well in order to ensure success. The CSF method is a very effective tool for providing information about customer needs and potential sources of competitive differentiation. Examples of CSFs include the development of a global sales network, superior new product development capabilities and the development of a best-in-class supply base.

The employment of these techniques in outsourcing evaluation allows valuable insights to be obtained into the types of capabilities that are likely to be a source of competitive advantage. This analysis will assist in identifying the major determinants of competitive advantage in the markets, the industries, or the strategic groups in which the organisation competes or might wish to compete. For example, General Motors Powertrain organisation have established 'customer clinics' in order to understand customer preferences and in particular the characteristics of the products for which customers are willing to pay a premium price (Fine *et al.*, 2001). These customer clinics involve eliciting the opinions of customers on performance characteristics such as fuel economy, acceleration, emissions, noise and vibration and relating these back to powertrain sub-systems. Workshops are conducted with engineering and manufacturing in order to relate engine and transmission sub-systems to the performance characteristics important to customers. This analysis allows General Motors to identify the drivers of customer preferences and the important processes that it should focus on. Focusing attention on the competitive environment and customer needs will involve applying the distinctive capabilities to meet these needs. It is crucial for a company to identify and define the critical skills that are central to its overall strategy. These critical skills will enable a company to differentiate itself from its competitors in the way in which it serves its customers. Maintaining control of the critical skills within the company will enable the company to sustain its competitive advantage in each product market. This process can be undertaken formally as part of the business planning process or on an ongoing basis in response to changing business priorities or customer requirements.

4.4.3   Stage 3 – relative capability analysis

Once the importance level of the activities under scrutiny has been identified, the next stage is concerned with analysing the capabilities of the organisation in these activities in relation to potential external sources – either competitors or suppliers. This will involve an analysis of the following.

- *Cost analysis* – an assessment of the relative cost position of the sourcing organisation in relation to both suppliers and competitors in the activities under scrutiny should be undertaken. The costs of sourcing the activity internally or from an external supplier should also be compared.

- *Evaluate the relevant activities* – each activity under scrutiny must be benchmarked against the capabilities of all potential external providers of that activity. This will enable the organisation to identify its relative performance for each activity along a number of selected measures.

In effect, this analysis is concerned with identifying the disparity between the sourcing organisation and potential external providers of the activities under scrutiny. This analysis allows the organisation to focus on whether it will be detrimental to their competitive position to outsource activities such as research and development, manufacturing, or assembly. A key issue in outsourcing evaluation is determining whether the organisation can achieve a sustainable competitive advantage by performing a critical activity internally on an on-going basis. Clearly, if the organisation can perform such an activity uniquely well, then this activity should continue to be carried out internally. However, many organisations assume that because they have always performed the activity internally, then it should remain that way. In many cases, closer analysis may reveal a significant disparity between their capabilities and those of the world best suppliers. This type of analysis is crucial to understanding the sustainability of a superior performance position. In the case of critical activities, such an understanding will have a fundamental influence on the sourcing option chosen due to the implications for the competitive position of the organisation. Understanding the source of the advantage will provide a sound indication of the difficulties with attempting to replicate or surpass superior performance levels in the activity. In order to undertake this analysis the following issues have to be considered.

- *The type of advantage* – this can be based on attributes such as lower costs, superior quality, service levels, etc. Superior performance in the activity under analysis can include a combination of these attributes.
- *The source of the advantage* – this is concerned with understanding how superior performance in the activity is achieved. In order to determine the source of the advantage, it is important to understand the relationships between activities, types of advantage, potential sources of advantage and resources as shown in Table 4.1.

This is a useful reference point for analysing the source of advantage in an activity. Consider a high-volume manufacturer that manufactures a number of standard products. In order to compete effectively, the organisation has focused primarily on the manufacturing process and outsourced support services such as inbound and outbound distribution and information technology. The manufacturer believes that it has a superior cost position in the manufacturing process relative to its direct competitors. The source of this advantage is based upon a number of factors including *scale economies* and *experience* in the process technologies

Table 4.1 *The relationship between activities, advantage and resources*

Sample activities	Types of advantage	Potential sources of advantage	Resources
• Operations • Distribution • Supply management • Marketing and sales • After-sales service • Product design and development • Human resource management	• Cost • Quality • Service • Time-to-market • Reliability	• Scale economies – for example, in production, marketing, etc. • Experience – for example, in production, design, marketing, service delivery, etc. • Location – for example, proximity to customers leads to higher service levels, etc. • Linkages – for example, internal cross-functional collaboration for rapid new product development, etc.	• Physical infrastructure • Equipment • People • Finance • Relationships • Reputation

associated with the manufacturing process. Furthermore, to create this advantage requires the deployment of a number of resources including equipment, people, finance and relationships.

The level of analysis at this stage will depend upon the importance level of the activity. Critical activities will require extensive analysis of performance levels in comparison to potential external sources. This analysis may involve a structured benchmarking approach to assessing the company's capabilities in the range of activities identified by top management in relation to the ability of potential suppliers and competitors to provide these activities. Using benchmarking will allow the company to look beyond the products to the operating and management skills that produce the products. It involves searching out the 'best of breed' of a process or skill, requiring active participation of the personnel who perform the function within the company.

4.4.4   Stage 4 – an analysis of the strategic sourcing options

This stage is concerned with evaluating the implications of the strategic sourcing options based upon the importance level and capability of the organisation in the activities under scrutiny in comparison to external sources. These two dimensions yield the potential strategic sourcing options as shown on the matrix in Stage 4 of the framework on Figure 4.1. The choice of these sourcing options should be considered in the context of the following determinants.

- *The disparity in performance* – this is concerned with understanding the sustain-ability of superior performance in the activity under scrutiny – either by the internal or external source. A clear understanding of the source of superior performance in the activity can be a reliable indicator of the sustainability of such a position. For example, where there is a significant disparity in perform-ance between the organisation and suppliers, it will be difficult to justify a substantial investment of resources in order to match or advance upon supplier performance.
- *Technology influences* – in many business environments, technology can influ-ence the choice of sourcing option. For example, advances in technology can rapidly erode any performance advantage an organisation may possess in a particular activity.
- *External considerations* – factors in the external environment such as the political context, market growth rates, the level of competition and barriers to entry should also be considered. For example, in a rapidly growing market with a high level of competitive rivalry, competitors will invest considerable resources to eliminate any superior performance advantage in an activity possessed by an organisation.
- *Behavioural considerations* – where outsourcing is being considered as an option, there are likely to be a number of behavioural issues that will affect the freedom of the organisation to outsource. For example, if outsourcing evaluation is handled insensitively, then there is likely to be significant workforce resistance to such a move particularly from employees that are going to be directly affected.
- *Supply market risk* – the level of risk associated with the relevant supply market should be analysed when outsourcing is being considered. There are a number of factors that indicate the level of supplier market risk including the number of suppliers and buyers in the supply market, the level of uncertainty and avail-ability of information on the general supply market or individual suppliers.

In many cases the inter-dependencies between each of these determinants will influence the choice of sourcing option. For example, consider a fast-moving consumer goods company that is considering whether it should manufacture and package internally one of its detergent products. The company focuses on enhancing its brand image as it believes it is an important factor for consumers when purchasing their products. It is unlikely that consumers are concerned with who manufactures and packages the detergent. Therefore, the company may consider outsourcing the manufacture and packaging to a supplier rather than operating its own plants. However, analysis of the supply market risk determinant may reveal that there are only a limited number of suppliers that possess the knowledge and scale in the manufacturing processes. These circumstances in relation to the supply market risk determinant may prevent the organisation from pursuing outsourcing as an option.

The implications of each quadrant on the matrix in Stage 4 of the framework on Figure 4.1 are now discussed in the context of these determinants.

Quadrant one
In this quadrant, there are external suppliers that are more capable than internally within the sourcing organisation for a critical activity. The sourcing organisation has the following options.

- *Invest to perform internally* – this option involves investing the necessary resources to bridge the disparity between the sourcing organisation and the more competent external providers of the activity. The selection of this option will depend upon the following:
  - *Significance of the disparity* – if the disparity is not significant then there is the potential to invest resources in order to perform the activity internally. For example, this option may be desirable in a case where the technologies involved in the activity are in the early stages of development and therefore may offer the potential for future growth.
  - *Type of disparity* – the type of disparity is crucial in determining whether it is feasible to invest the necessary resources to match the superior performance of external sources. For example, analysis of the activity may reveal that the disparity in performance is in an area such as quality or productivity, which can be addressed through an internal improvement initiative.
- *Strategic outsource* – this option is likely if the analysis has revealed that it is not possible to attempt to bridge the disparity in performance. The organisation may consider outsourcing such an activity, which is likely to diminish in importance in order to focus resource and effort on activities that promise to be a source of competitive differentiation in the future. Moreover, in certain circumstances the organisation may have no choice other than outsourcing because of internal capacity constraints. Strategic outsourcing is most appropriate when the organisation feels the advantage the external source has in the activity is too difficult to replicate.
- *Acquisition* – another potential option is to acquire an organisation that possesses superior performance in the activity. For example, the sourcing organisation may have decided that it cannot achieve the required performance levels internally and there is a high level of risk associated with using an external supplier.

Quadrant two
In this quadrant, the organisation is more competent than any other potential external sources in a critical activity. In this case, the sourcing organisation has the following options.

- *Perform internally and develop* – again, as with quadrant one it is important to consider both the significance and type of disparity in performance in the activity. For example, if the sourcing organisation has built up a significant performance advantage through scale economies over time, then it is going to be extremely difficult for external sources to replicate such a capability. Clearly, keeping the activity internal is the most appropriate when the sourcing organisation is in a strong position to sustain its performance advantage over time.

- *Strategic outsource* – ideally, an organisation wants to have superior performance in as many critical activities as possible. However, it is only possible to possess leadership in a limited number of activities due to the resources required to maintain such a position. For example, the sourcing organisation may consider that it is not able to sustain superior performance in this activity in the future and therefore may decide to outsource the activity to the most competent external source.

Quadrant three

In this quadrant, there are external suppliers that are more capable than internally within the sourcing organisation for an activity not critical to business success. These activities are suitable candidates for outsourcing. The most significant influences are the level of supply market risk and the constraints that impact upon the freedom of the organisation to outsource. For example, if there are only a limited number of capable suppliers in the supply market, the sourcing organisation may decide to continue to perform the activity internally. In addition, internal constraints such as the threat of industrial action may impinge upon the freedom to outsource.

Quadrant four

In this quadrant, the organisation is more competent than potential external sources in an activity that is not critical to business success. Although the sourcing organisation is more competent than external sources, the activity is not central to competitive advantage. Therefore, the organisation should consider outsourcing such an activity and focusing resources on building capabilities in activities that are more critical to the success of the organisation. However, if the organisation decides to outsource then it will have to develop the capabilities of a supplier to the level achieved internally. There are a number ways of achieving this including a supplier development programme, through the transfer of employees and equipment to a suitable supplier or a management buy-out of the activity under consideration.

4.4.5   Stage 5 – relationship strategy

As organisations increasingly outsource more important activities, they have been attempting to develop collaborative relationships with suppliers as they seek to reduce the risks associated with outsourcing. By employing astute relationship management mechanisms it is possible for organisations to leverage the capabilities of suppliers in a more effective way than competing organisations. Organisations are now beginning to recognise the benefits of adopting approaches for managing across and beyond the boundaries of the organisation. There is a growing awareness that competitive advantage can be based as much upon managing beyond the ownership boundaries of the organisation rather than on management within those boundaries. This stage outlines the steps involved in building relationships across organisational boundaries in order to achieve the objectives established for outsourcing. The key steps in this stage include establishing the objectives for outsourcing and selecting the appropriate supply relationship to achieve these objectives.

Establish outsourcing objectives

Establishing clear objectives will assist in the implementation of a successful outsourcing strategy. Outsourcing objectives can include reducing the costs of sourcing the activity, enhancing quality levels and obtaining higher levels of service in the provision of activity. The objectives for outsourcing are important from a number of perspectives including the following.

- *Selecting the supply relationship* – the objectives established for outsourcing will assist in selecting the type of supply relationship that should be adopted. For example, if the organisation is focusing primarily on attaining the lowest price, then an adversarial relationship is likely to be the most appropriate supply relationship.
- *Monitoring supplier performance* – the outsourcing objectives will inform the metrics used to assess supplier performance throughout the life of the relationship.
- *Monitoring the nature of the supply relationship* – it is important to assess whether the relationship with the supplier is meeting the overall outsourcing objectives. This will involve analysing issues such as the level of dependency and the depth of collaboration in the relationship.
- *The competitive position of the sourcing organisation* – the outsourcing objectives should be linked with the strategic goals of the sourcing organisation. For example, the development of long-term collaborative relationships with suppliers can contribute to the overall business strategy of the organisation.

Supply relationship strategy

The choice of the relationship with the supplier will be influenced by the objectives for outsourcing the activity. For example, in outsourcing an activity that is not critical to business success, the objectives are likely to be tangible and include factors such as reduced costs, better service in the form of on-time deliveries, inventory reduction and better quality levels. In this case, an adversarial relationship will be more appropriate particularly if there are a significant number of capable suppliers in the supply market. In certain circumstances, the buying organisation will gain more by pursuing a relationship where it possesses the balance of power rather than pursuing a relationship based upon equality and the mutual sharing of benefits. The buyer can achieve the benefits of collaboration whilst still maintaining the balance of power in the relationship. In other words, in order to achieve the most beneficial supply arrangement the organisation may pursue a buyer–supplier relationship – ranging from close collaboration to adversarial – which will enable it to maximise competitive advantage. Selection of the appropriate relationship strategy is influenced by the following.

- *The importance of the activity* – the importance of the activity being outsourced is a reliable indicator of the attention that the buyer should give to managing the relationship with the supplier.
- *Supply market risk* – includes factors in the supply market that can create difficulties for the buyer in managing the relationship with the supplier. For example, a high level of risk in the supply market will necessitate the development of a supply strategy to secure long-term supply and the employment of mechanisms to encourage the supplier to make investments that are specific to the needs of the buyer.

These two dimensions yield the potential relationship strategies as represented on the matrix in Stage 5 of the framework in Figure 4.1. The implications of each of the quadrants on the matrix are now considered.

Quadrant one

In this quadrant, the sourcing organisation has two options in terms of relationship strategy.

- *Competitive collaborative* – this strategy is most appropriate when the supply market is highly competitive with many suppliers competing for business. Suppliers have limited bargaining power and customers can easily switch to alternative sources of supply. Therefore, the buyer can develop a relationship with a supplier for as long as that supplier maintains a position of leadership in the activity concerned.
- *Close collaboration* – rather than have a number of suppliers competing against each other for the business, it is possible to establish a close long-term strategic collaborative relationship with one of these suppliers.

Quadrant two

In this quadrant, the most appropriate relationship is close collaboration. Close collaboration involves the adoption of a long-term collaborative buyer–supplier relationship. This approach is suitable in the case of a critical activity for which there are only a limited number of suppliers in the supply market. Adopting this approach allows the buying organisation to establish and build a mutually advantageous relationship with the supplier. Through specialised investments in the relationship the buying organisation and the supplier can develop a relationship that delivers benefits that are unavailable to competitors. The presence of relation-specific investments such as collaboration in new product development, the development of inter-organisational information systems and integrated order management and delivery systems can create a high level of mutual dependency between the buying organisation and the supplier.

Quadrant three

In this quadrant, it is possible to source the activity from a wide number of suppliers in the supply market. The supply market is extremely competitive with many suppliers aggressively competing for business from customers. The most appropriate strategy in this case is to source the activity from the most competitive supplier in terms of price, delivery and quality. Moreover, the buyer should be aware of the attempts of suppliers to build any dependency into the relationship. For example, developments in the supply market such as acquisitions by suppliers can affect the balance of power in the relationship.

Quadrant four

In this quadrant, there are factors in the supply market, which can create uncertainty and vulnerability in supply. For example, the supply market may be highly concentrated with a few large suppliers. Careful attention should be given to ensuring continuity of supply. The buyer may pre-empt any supply problems by either establishing a long-term relationship with the supplier or maintaining some capability internally in the activity being outsourced. It is possible that the sourcing organisation will have outsourced a number of activities to a single supplier. In addition, these outsourced activities may have differing levels of importance. In this case, the sourcing organisation will choose the appropriate supply relationship on the basis of the amount of business it has with the supplier rather than on the importance of a single outsourced activity. For example, a close collaborative relationship is appropriate in a situation in which the sourcing organisation's expenditure is a significant ratio of the supplier's total sales revenue.

4.4.6   Stage 6 – establish, manage and evaluate the relationship

This stage is concerned with implementing the outsourcing strategy. There are a number of issues that have to be addressed at this stage.
- *Supplier selection* – the supplier selection strategy must enable the organisation to achieve the outsourcing objectives. Supplier selection has become

increasingly important as organisations outsource more important activities and develop more collaborative and longer-term relationships with key suppliers. The starting point in supplier selection involves the buyer determining what is required from the supplier. This will involve analysing the activity being outsourced in order to determine the skills and resources required to deliver the activity. Once the sourcing organisation has established what is required from the supplier, it can then determine the criteria to evaluate potential suppliers. The type of relationship being adopted will influence the criteria chosen. For example, the predominant features of an adversarial relationship include an emphasis on price followed by quality, delivery and service. Consequently, these features will dominate the criteria used in the evaluation of potential suppliers. In the final selection decision, potential suppliers will then be evaluated against their ability to meet these criteria.

- *Contracting issues* – organisations considering outsourcing must have an understanding of the business and legal issues associated with outsourcing and how some of these issues can be dealt with in a contract. An outsourcing contract will include a number of aspects such as a service level agreement (SLA), the transfer of staff and assets, price and payment terms, liability and contract termination. The type of activity being outsourced and the level of risk in the supply market will influence the design of the contract. For example, in the case of a straightforward activity with a significant number of suppliers in the supply market, the objectives of outsourcing the activity and requirements of the supplier are well defined and can be encapsulated within the contract. However, in some cases it is not possible to design a contract, which accounts for all future contingencies. In such circumstances, it is more appropriate to develop a relationship that fosters collaboration to compensate for any gaps in the contract. For example, relational mechanisms such as a joint problem-solving culture can be employed to deal with any deficiencies in performance on the part of the supplier not allowed for in the contract.

- *Managing the relationship* – once the activity is outsourced, the direction and management of the buyer–supplier relationship will be influenced by the objectives of the outsourcing strategy. The success of the relationship will be largely determined by how it is managed at the operational level. Therefore, it is critical that both the buyer and supplier have the necessary skills and resource to manage the interaction process at the operational level. For example, the interaction patterns between the relevant participants at the buyer–supplier interface are critical in ensuring the development of a close collaborative relationship. This will involve a significant resource commitment and will also rely on input from senior management at various stages in the development of the relationship.

- *Relationship performance evaluation* – the sourcing organisation must have a formal mechanism to determine whether the supplier is meeting the performance levels set and whether the objectives in its approach to relationship management are being achieved. There are a number of aspects to this evaluation including the following:

- *Supplier performance* – this is concerned with determining whether the supplier is delivering to the required standards during the contract. This analysis will focus on performance metrics related to quality, delivery, service and ability to reduce costs.
- *The strength of the relationship* – evaluation of the strength of the supply relationship will be guided by the initial objectives established for the out-sourcing process. This evaluation is of vital importance in the case of colla-borative relationships. For example, monitoring the strength of a close collaborative relationship will involve analysing factors such as joint problem solving, high levels of information exchange and top management commit-ment from both the buyer and supplier.
- *The level of dependency* – changes in the internal and external environment can have a positive or negative effect upon the level of dependency for the buyer over the lifetime of the relationship. For example, over time the importance of the outsourced activity can increase or diminish due to changes in the external environment such as advances in technology. Changes in factors such as the number of available suppliers, competitive demand for supply and the performance of other suppliers also can impact the level of dependency in the relationship.

This evaluation serves as a context for action, which can involve maintaining the relationship at its current level, further developing the relationship or discontinu-ing or reducing the scope of the relationship.

4.5   Implications for theory

The framework for outsourcing evaluation and management has illustrated the interdependencies of certain aspects of the theoretical influences introduced in the literature discussion in the previous chapter. The industry view has influenced the analysis of the importance of organisational activities. In particular, Porter's (1985) generic drivers of competitive advantage have influenced the analysis of the importance of organisational activities. Activities that underpin the achievement of a generic driver are critical for organisational success. The capability analysis stage is closely related to the resource-based view. Analysing organisational capability can assist in identifying 'valuable strategic' activities that are a source of competitive advantage. Furthermore, the analysis carried out in relation to the importance and capability of the organisation in certain activities can be self-reinforcing. For example, the importance of an activity may be further emphasised if the capability analysis reveals that the organisation is much more competent than any suppliers or indeed competitors. Also, in certain circumstances the order of analysis in relation to stages two and three can be reversed. For example, the impetus for considering the suitability of certain activities for outsourcing may have come from poor internal performance. In this case, the organisation may firstly undertake a capability analysis exercise both to determine the causes of poor performance and the capabilities of suppliers and competitors. If no scope for internal improvement is identified, then the organisation will have to consider the impact of the activity on its competitive position (i.e. importance analysis) before deciding whether outsourcing is appropriate. Whether the analysis focuses on the internal or external dimension, the organisational activities that underpin the drivers of competitive advantage should be determined. Clearly, activities that underpin a generic driver and in which the organisation has a strong position will be deemed critical and have to be controlled and developed internally. Alternatively, an activity that is not critical to competitive advantage may be deemed suitable for outsourcing.

Combining the activity importance and capability analysis stages offers a number of potential sourcing options as shown in Stage 4 of Figure 4.1. In relation to quadrants one and two, the choice of sourcing option is influenced primarily by the sustainability of any superior performance position held by either the sourcing organisation or its competitors and/or suppliers. The analysis employed at this stage is closely related to the resource-based view. Barney's (2002) four criteria including *value, rarity, imitability,* and *organisation* can assist in the evaluation of the sustainability of any performance superiority. For example, the *Invest to Perform Internally* option in quadrant one is most likely when the sourcing organisation can easily replicate a superior performance position held by its competitors or suppliers. Alternatively, the *Perform internally and develop* option in quadrant two is most appropriate when the sourcing organisation possesses a capability that is built on rare, valuable, non-substitutable and difficult to imitate resources (Barney, 2002). The *Strategic outsource* option in both quadrants one and two – i.e. outsourcing a critical activity through a close collaborative relation- ship – is likely in a rapidly changing environment where speed and flexibility are particularly important for competitive advantage. Organisations can benefit from sourcing critical activities from more capable suppliers. For example, in a manu- facturing context, outsourcing a critical activity can spread the company's risk in the area of technology developments among a number of suppliers. The buyer is not limited to its own internal capabilities but can access new product ideas that it could not possibly hope to generate itself. In relation to quadrants three and four, activities located in these quadrants are not critical to competitive advantage and therefore should be outsourced if a capable supply market exists.

Where outsourcing is considered a suitable option, the next stage involves selecting the most appropriate supply relationship. This analysis is influenced by both the importance of the activity and supply market conditions as shown in Stage 5 of Figure 4.1. Both the TCE and RV have influenced the relationship strategy options. The choice of an *Adversarial* relationship in quadrant three is comparable to market governance in TCE. In quadrant four, the *Secure supply* option is also influenced by TCE which is essentially concerned with employing complex contractual safeguards to reduce the level of risk in the relationship. For example, contractual safeguards can be employed to prevent either party behaving opportunistically when there is a high level of asset specificity in the relationship. The presence of uncertainty can also create difficulties in the relationship because either the buyer or supplier may have to adapt in response to unexpected changes. However, in some cases, the employment of complex contractual safeguards cannot guard against supply market risk in the form of opportunism and uncertainty. This situation is reflected in the *Close collaboration* in quadrant two which is influenced by the relational view. For example, the sourcing organisation can employ relational and alliance building approaches in order to strengthen the relationship with the supplier and reduce the associated risks with outsourcing in less contractual and legalistic ways. Inter-organisational collaboration can also be employed to access and develop complementary resources that are required for business success.

The framework does not use the core/non-core language in the evaluation and management of the outsourcing process because of the common misconceptions amongst both practitioners and academics. Much of this confusion emanates from a lack of understanding of the core competence concept and its relationship with outsourcing. The framework employs terminology to create a clearer understanding of both the evaluation and implications of outsourcing. For example, instead of using the term 'non-core' to describe outsourced activities, the framework emphasises that critical activities can be outsourced and the way in which the supply relationship is managed must reflect their importance. Therefore, the outsourcing framework emphasises that outsourcing can be used to build and strengthen capabilities across organisational boundaries. There is also support for this view from Hamel and Prahalad (1994) who argue that a core competence is a bundle of skills and technologies rather than a single discrete skill or technology. Indeed, they emphasise the importance of integration both internally and externally across organisational boundaries, suggesting that a core competence may be viewed as communication, involvement, and a deep commitment to working across organisational boundaries involving the interaction of many levels of people and functions. This is supported by John Kay (1995) who argues that an organisation's distinctive capability may be based upon innovation, reputation, or the organisation's architecture, the system of relationships within the firm and between firm, suppliers and customers. The relationships that provide architecture lead to competitive advantage through the acquisition of organisational knowledge, the establishment of organisational routines and the development of a co-operative ethic. Such capabilities require a seamless integration of activities, their operation and development.

Referring back to the discussion in the literature in the previous chapter on governance skills, it can be seen that the framework emphasises the importance of governance skills both internally and externally. If an organisation possesses effective governance skills, it has access to a broader range of resources to choose from. Access to a broader range of resources through superior governance skills can be a source of competitive advantage. These governance skills in managing external suppliers can be employed in both a collaborative and adversarial fashion in the appropriate context. Much of the discussion so far has centred on the potential benefits of pursuing collaborative relationships with suppliers. However, in some contexts, as the large UK supermarket retailers Tesco and Sainsburys have shown, the use of adversarial relations with suppliers can be an extremely powerful vehicle for profit maximisation. Therefore, skilful governance is a type of capability in itself. In fact, governance can be the most critical activity that an organisation undertakes. For example, Marshall (2001) has found that in rapidly changing industries, identifying the importance of the activity and then following a prescribed path appropriate to the level of importance can be an unsuitable way of evaluating and managing the outsourcing process. In this context, it is better for organisations to direct more attention to scanning and analysing the supply market to identify capable suppliers that can enhance their competitive position. The challenge for many organisations is in managing beyond the ownership boundary than on managing within it. The contribution of outsourcing to the achievement of competitive advantage is heavily dependent upon how effective the organisation is in leveraging the capabilities of suppliers relative to its competitors.

Illustration 4.2

A comparison of outsourcing in the public and private sectors

Bernard Burnes and Antisthenis Anastasiadis conducted a study comparing the approaches and experiences with outsourcing in both the public and private sectors. They carried out two case studies of organisations that had extensive experience of outsourcing. The public sector organisation chosen was one of the largest police forces in the UK; the private sector organisation was an insurance broker providing specialist risk management, advisory and other services to a range of corporate and institutional clients on a global basis. Their analysis was undertaken along the dimensions of the decision to outsource, supplier selection, contractual arrangements and managing the relationship. They found similarities in the experiences of the organisations studied including the objectives of achieving value for money and experiencing some of the pitfalls associated with outsourcing. However, the major contrast between the two organisations was in the reasons and motives for outsourcing. These similarities and differences were found across these dimensions of analysis.

- *Decision to outsource* – the public sector organisation was compelled by legislation to put certain activities out to tender. If it could be demonstrated that an activity could be performed better by an external organisation, then the activity should be put out to tender. In contrast, the insurance organisation's policy was that it would only be involved in direct service delivery if more value could be added than from outsourcing the activity. It was not obliged to undertake an analysis of the suitability of all business activities for outsourcing purposes.
- *Supplier selection* – again under legislation the public sector organisation had to carry out a rigorous and open supplier selection process. The tender had to be advertised across the EU and the lowest bidder chosen providing the tender specifications could be met. In the event of supplier failure, the total tendering process had to be gone through again. In the case of the insurance company, it could use what ever method it chose to select the supplier. For example, the organisation could identify the most capable suppliers and negotiate with them in order to determine who would offer the best service. Once the supplier was selected further negotiations could take place in order to adapt to any changes. This contrasts with the public sector. When a contract specification has been agreed, it was extremely difficult for the contractor or supplier to change it.
- *Contractual arrangements* – the analysis revealed similarities in both contract length and specifications in the public and private sector organisations. However, in the public sector, longer and more formal contracts were offered due mainly to the larger scale of the contracts. In relation to performance, both organisations used service level agreements (SLAs) to define their requirements for the outsourced activity. Gain-share clauses were used by the private sector organisation. Gain-share clauses are employed to encourage suppliers to achieve improvements that benefit both the customer and supplier. Gain-share clauses were less prevalent in the public sector because of the emphasis on achieving value for money. If it was found that some suppliers were achieving gains beyond those established in the contract, this might be interpreted as corruption or incompetence on the part of the public sector organisation. In relation to conflict resolution, the public sector organisation appeared to pursue best practice because of the imposed legal framework and greater public scrutiny. Private sector organisations tended

to resolve conflict either through inter-personal mechanisms or seek redress through the courts.

- *Managing the relationship* – both organisations attempted to adopt a collaborative approach. Although the public sector organisation adopted elements of collaboration, the use of competitive tendering mechanisms was more closely related to the characteristics of adversarial relationships. The private sector organisation promoted collaborative relations with its key suppliers and had a more positive and strategic approach to outsourcing. Both organisations had different practices in relation to supplier development. In the public sector organisation the legislative framework significantly inhibited the employment of supplier development. For example, working with suppliers to achieve improvements might create the impression of favouritism particularly if it gives the supplier an advantage in the re-tendering process. In contrast, in the private sector organisation, suppliers were encouraged to improve service levels both independently and jointly with the customer.

In conclusion, the findings showed that although outsourcing is becoming more prevalent in both the public and private sectors, neither has a monopoly on best practice. The private sector could benefit from the more structured approach adopted in the public sector, whilst the public sector could benefit from the strategic focus of the private sector.

Source: Burnes, B. and Anastasiadis, A. (2003). Outsourcing: a public–private sector comparison. *Supply Chain Management: An International Journal*, 8, 4, 355–366.

REFERENCES

Barney, J. A. (2002). *Gaining and Substaining Competitive Advantage*. New Jersey: Prentice Hall, Second edition.

Barthelemy, J. (2003). The seven deadly sins of outsourcing.. *The Academy of Management Executive*, 17, No.2, 2, 87–98.

Bowman, C. (1998). *Strategy in Practice*. London: Prentice-Hall.

Bryce, D. J. and Usseem, M. (1998). The impact of corporate outsourcing on company value. *The European Management Journal*, 16, No. 6, 635–43.

Fine, C. H., Vardan, R., Pethick, R. and El-Hout, J. (2001). Moving a slow-clockspeed business into the fast lane: strategic sourcing lessons from value chain redesign in the automotive industry. http://mitsloan.mit.edu/research/clockspeed/finevaluechain.doc

Greenberg, E. R. and Canzoneri, C. (1997). *Outsourcing: The AMA Survey*. New York: The American Management Association.

Hamel, G., and Prahalad, C. K. (1994). *Competing for the Future*. Boston: Harvard Business Press.

Hogwood, B. W and Gunn, L. A. (1984). *Policy Analysis for the Real World*, New York: Oxford University Press.

Kaplan, R. S. and Norton, D. P. (1996). Using the balanced scorecard as a strategic management system. *Harvard Business Review*, 74, No. 1, 75–85.

Kay, J. (1995). *Foundations of Corporate Success*. Oxford: Oxford University Press.

Lebas, M. (1999). Which ABC? Accounting based on causality rather than activity-based costing. *European Management Journal*, **17**, No. 5, 501–11.

Linder, J. C., Cole, M. I. and Jacobsen, A. L. (2002). Business transformation through outsourcing. *Strategy and Leadership*, **30**, No. 4, 23–28.

Lonsdale, C. and Cox, A. (1997). Outsourcing: risks and rewards. *Supply Management*, 3 July, 32–4.

Marshall, D. (2001) *The Outsourcing Process: From Decision to Relationship Management*. Ph.D. Thesis, University of Bath.

McIvor, R. (2000). A practical framework for understanding the outsourcing process. *Supply Chain Management: An International Journal*, **5**, No. 1, 22–36.

Normann, R. and Ramirez, R. (1993). From value chain to value constellation: designing interactive strategy. *Harvard Business Review*, **71**, No. 4, 65–77.

Porter, M. E. (1980). *Competitive Strategy: Techniques for Analysing Industries and Competitors*. New York: The Free Press.

Porter, M. E. (1985). *Competitive Advantage*. New York: Free Press.

Porter, M. E. (1991). Towards a dynamic theory of strategy. *Strategic Management Journal*, **12**, 95–117.

Probert, D. R. (1996). The practical development of a make or buy strategy: the issue of process positioning. *Integrated Manufacturing Systems*, **7**, No. 2, 44–51.

Saunders, M. (1997). *Strategic Purchasing and Supply Chain Management*. London: Pitman Publishing.

Stabell, C. B and Fjeldstad, O. D. (1998). Configuring value for competitive advantage: on chains, shops, and networks. *Strategic Management Journal*, **19**, 413–437.

5 Determining the current boundary of the organisation

5.1 Introduction

This stage in outsourcing evaluation and management is concerned with identifying the major activities that have to be performed in order to create and deliver the range of products and services offered by the organisation to its customers. This involves identifying the major activities that have to be performed internally as well as those performed by external sources. This will provide an outline of the scope of the organisation both upstream into the supply chain and downstream towards end customers of the products and services. In effect, it is concerned with identifying the current boundary of the organisation. Additionally, as will be shown in the following chapters, segmenting the organisation into activities allows an organisation to identify the activities that are a source of competitive advantage, activities in which it is equal to competitors and activities in which it is weaker in relation to competitors. In order to perform this analysis, the resources owned and deployed by the organisation must be identified as well as the activities that have to be carried out so that the organisation fulfils the needs of its customers. A number of frameworks appropriate for activity analysis are presented in this chapter including the value chain, value shop and value network. Analysing the organisation in terms of activities allows outsourcing evaluation to be carried out in a number of contexts including:

- an organisation may undertake an evaluation of its entire business and identify potential opportunities for outsourcing from this analysis;
- an organisation may select certain parts of the business that it considers as suitable candidates for outsourcing;
- an organisation may consider outsourcing as an appropriate means of improving performance in certain activities that have been causing problems.

In each of these contexts, the organisation must have a means of segmenting various parts of the business into activities. Segmenting the organisation into activities serves as a basis for determining the importance level and capability of the organisation in the relevant activities. The very existence of an organisation is often determined by whether the organisation performs certain activities internally rather than source them from external sources. Such decisions on the boundary of the organisation can include on the one extreme, the procurement of stationery, while on the other, the internal production of a critical part of the end product. Previously, in many organisations there was a focus on participating in or owning as many stages as possible in the industry chain – vertical integration. Traditionally, large organisations have attempted to have as many activities as possible within the boundary of the organisation. The arguments in favour of such a strategy were based on the premise that the organisation had the potential to exercise greater control of the strategic resources that would enhance its competitive position. These resources can be located upstream in the supply chain or downstream towards the end-customer. For example, an organisation may acquire a supplier when the item being purchased is a major cost component of the end product and cannot be produced in-house competitively. Additionally, an organisation may acquire a network of retailers in order to build a committed group of sales outlets promoting and selling its products to end-customers. However, as the business environment becomes more complex and unpredictable, there are significant risks and problems with pursuing such a strategy including the following (Alexander, 1997)

- *Cost of ownership* – in periods of scarce resources there is more of a focus on understanding the opportunity costs of having capital sunk into a collection of assets. For example, such assets can be a significant drain on management time and cash flow. These resources can be employed on more productive and innovative activities that create greater value for the customers of the organisation.
- *Ineffectiveness of ownership* – it has become increasingly recognised that many large highly integrated organisations have under performed. There is no guarantee that owning a number of stages in the industry chain will create competitive advantage. The influence of bargaining power in a single part of the industry value chain can be the dominant source of value for the customer. For example, it is possible for a company to capture most of the value in an industry by integrating the component and service elements – performed entirely by others – of the product and delivering it to the end-customer.
- *Inflexibility of ownership* – in times of rapid change and uncertainty the ownership of an asset can be a considerable risk. During the 1980s and 1990s, many vertically integrated companies have found vertical integration to be competitively inflexible in a rapidly changing business environment. Physical assets such as manufacturing facilities or equipment can rapidly become obsolete. A high level of integration in the industry value chain can also be inflexible with organisations owning assets that are of decreasing value.

Organisations that are too broad in their scope of activities often lack strategic direction and have no strategic basis on why certain activities are performed internally. In many cases, this can be as a result of cultural or historical ties to

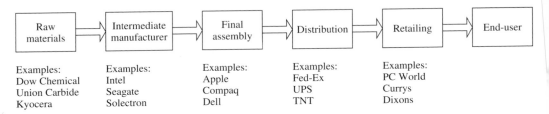

Figure 5.1 Stages in the personal computer industry chain

the activity rather than the activity making a contribution to competitive advantage or the organisation possessing a distinctive competence in the activity. Furthermore, many organisations are not accustomed to employing external suppliers to provide a wide range of activities. For example, this is a problem for public sector organisations in particular, that lack the management skills and organisational structure to rapidly access the capabilities of product and service providers in supply markets. Public sector organisations tend to be risk averse and do not possess the mindset or culture to fully avail of the value that can be obtained in supply markets. Moreover, some organisations can further exacerbate the problem of organisational scope by attempting to take advantage of opportunities in the external business environment without fully understanding the resource implications. These same organisations often lack the necessary resource to deliver on their current portfolio of activities. Further increasing the scope of an already over-committed organisation leads to a lack of focus and direction. Crucially, this lack of direction will also manifest itself at the level of the employee, who is likely to be stretched in a number of directions and as a consequence lacks the necessary focus to achieve the required levels of performance in their specialist area.

These risks and drawbacks have led many organisations to pursue a strategy of vertical disintegration. It is most unusual for an organisation to carry out or own all the activities in the overall industry value chain. Rather than participating in as many activities in the industry chain as possible, organisations have been attempting to specialise and compete in specific activities. The basis for carrying out an activity is to add value to the product. For example, in the case of a series of activities in the manufacture of a product, a company takes a product produced in the previous stage and transforms it in a way that it is worth more to a company in the next stage in the chain. Using the personal computer industry as an example illustrates this concept. Figure 5.1 shows the stages involved in the creation and delivery of a personal computer to the end customer.

Raw materials companies include the manufacture of speciality ceramics, chemicals and metals. For example, Kyocera makes the ceramic substance for semiconductors. The next stage in the industry is the manufacture of components that include these raw materials. For example, manufacturers such as Intel, Seagate and Solectron transform these raw materials into the major components such as microprocessors, disk drives and motherboards. At the next stage, assembly companies such as Apple and Packard-Bell specialise in the purchase of these major components and assembly of them into personal computers. They are adding value to the components they purchase. Many of these personal computers are then sold to distributors or retailers who then sell them to end customers. These distributors specialise in the activity of making the product accessible to customers and providing service and support. In effect, each company has made a decision to operate and compete in a particular part of the industry chain. Each company specialises in this particular part of the chain in order to add value. Most of these companies specialise on one part of the chain and have not integrated either upstream or downstream into adjacent stages.

Illustration 5.1

Value chain positioning in the chemical industry

Robert Collins and Kimberly Bechler carried out a study of outsourcing strategies in both the chemicals and automotive industries. This study was based on work with a number of leading companies including BP Chemicals, DuPont de Nemours, Exxon Chemical International, GKN Automotive, and Volkswagen Audi AG. In particular, the study of the chemical industry provides an illustration of the motives behind companies positioning themselves in a particular part of the industry value chain and their approach to outsourcing. The application and development of chemical compounds can be in a range of diverse areas including welding, flavours, fragrances, dyes and oilfield and industrial chemicals. The nature of these applications and end-user needs influences production volumes. For example, in the case of industrial chemicals global demand can be as much as hundreds of thousands of tonnes, whilst in the case of the fragrances used in perfumes the amount can be in grams. The choice by the chemical firm to compete in a particular product area will influence both its position in the industry value chain and whether it outsources manufacturing as shown in Figure 5.2. For example, BP Chemicals, a bulk chemicals manufacturer, does not outsource manufacturing activities but rather invests in the capabilities required to produce hydrocarbon products to its customers. Alternatively, DuPont de Nemours, in its agricultural crop protection business, outsources the manufacture of chemical intermediaries that comprise the development of complex molecules that are used in pesticides, herbicides and fungicides.

BP Chemicals
BP Chemicals competes in the bulk chemical industry. It specialises in the production of petrochemical derivatives and is positioned on the primary

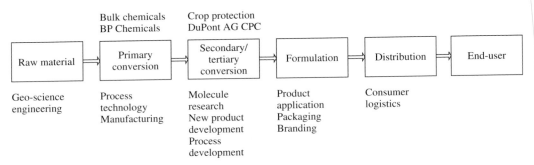

Figure 5.2 The chemical industry value chain. From the Institute for Supply Management.[TM] Reproduced with permission.

conversion stage of the industry value chain as shown in Figure 5.2. Organisations that compete in this part of the industry value chain focus on process efficiency and process technology development. These organisations create value for their customers by developing and applying a particular process technology. The goal of pursuing developments in process technology is to maximise plant capacity and achieve economies of scale. In order to achieve competitive advantage, BP Chemicals focuses on achieving increased process efficiencies and process innovations. Therefore, it does not outsource manufacturing as it considers it has a distinctive capability in the process technologies associated with manufacturing that allow it to provide a cost competitive product to its customers.

DuPont agricultural crop protection chemicals
DuPont participates in the agricultural crop protection business. This part of the industry is bio-science-based characterised by high levels of commitment to research and development. Moreover, new product development can be an extremely lengthy and costly process. However, strong capabilities in research and development in areas such as reducing time to market and developing product innovations are regarded as essential for achieving competitive advantage. In crop protection, DuPont is one of the leading companies in providing application technology, products and services to agriculture. The company is positioned in the secondary/tertiary conversion stage of the industry value chain where capabilities in molecular research, new product development and process development are required. For DuPont competitive advantage is based on achieving timely and efficient product innovations for customers. Research and development focuses on the design of a molecule having the chemical and physical properties that match the intended application. This can be an extremely complex process that involves the screening of thousands of variants. In production, manufacturing progresses from laboratory-scale production, to pilot plant

and then full volume production. Plant and equipment for production may require an investment of up to $100 million. The time span from initial screening to product launch can range from 5 to 10 years. Therefore, DuPont focuses on research and development and outsources manufacturing to specialists that have the capacity and the potential to achieve the required economies of scale. The company can access these capabilities without having to commit the necessary capital investment required for manufacturing.

Source: Collins, R. and Bechler, K. (1999). Outsourcing in the chemical and automotive industries: choice or competitive imperative. *The Journal of Supply Chain Management*, **35**, No. 4, 4–11.

5.2   Value chain analysis

Value chain analysis provides a framework for segmenting an organisation into a number of activities and identifying organisational capability in these activities. A useful starting point in segmenting the organisation into a number of activities is to identify the resources owned by the organisation. In order to perform activities that create competitive advantage, organisations need to access resources. The resources of a typical organisation can be grouped into the following categories.

- *Tangible resources* – include resources that can be quantified and observed. Examples of tangible resources include physical and financial resources. Physical resources include plant, equipment, buildings, and location. It is unusual for the physical assets of an organisation to be a source of sustainable competitive advantage. Most physical assets such as factories and equipment can be readily purchased on the open market. Furthermore, physical resources can rapidly become obsolete and a competitive burden for an organisation. For example, in high technology industries there is considerable risk associated with having obsolete technology in the form of expensive equipment that can become outdated and uncompetitive. Financial resources can encompass obtaining capital, managing cash, debt and credit control and the management of the relationships with the banks and other credit agencies. Financial resources are rarely a source of competitive advantage due to the relative ease with which organisations can acquire capital from credit agencies such as banks, stock markets and venture capitalists. However, there are instances where financial resources can be used as a source of competitive strength in an industry. For example, companies with large cash reserves have an advantage in the event of a price war or a recession. Moreover, organisations with considerable financial resources may be in a better position to attract high calibre personnel in high profile industries where key individuals can have a major impact upon organisation performance.

- *Intangible resources* – include resources that are deeply embedded in an organisation's systems and have been developed over time. Examples of intangible resources include brand names, reputation, and technological and marketing know-how and human resources. These resources can be of considerable value. For example, companies invest considerably in building a strong brand for their products that will encourage customers to buy their product as an automatic choice. Reputation is another intangible resource that can be of value to the organisation and its customers. For example, organisations can create a reputation through the provision of reliable, well-designed products and good customer service. Human resources encompass management skills, experience and culture. In particular, the culture of people within an organisation can have significant influence on organisational success. In industries where there is rapid change an adaptive and dynamic culture can be a crucial source of competitive advantage. Furthermore, in professional-service organisations human resources are critical to the success of the organisation contributing directly to its performance and image. Increasingly, knowledge held in human resources is seen as a significant differentiator for organisations. Organisations are adopting a range of strategies to better leverage the knowledge capabilities of their staff including the creation of mechanisms to facilitate the collection, recording and dissemination of that knowledge. The creation of an intangible resource can depend upon cross-functional integration and collaboration both internally and externally. In many circumstances, an intangible resource such as reputation or brand can be a major source of competitive advantage.

In the resource-based view literature there are various terms and definitions to explain the concept of resources. Wernerfelt (1984) was one of the first authors to use the term resource in the context of strategy. Firms should exploit their existing resource base and develop new resources in order to establish a sustainable competitive position or 'a differential resource position'. Barney (2002) describes resources as all assets, capabilities, competencies, organisational processes, firm attributes, information and knowledge that are controlled by the firm. However, some authors distinguish between resources and capabilities. Amit and Schoemaker (1993) define resources as stocks of available factors owned or controlled by the firm. Capabilities are specific to the firm built up over time through the co-ordination and integration of resources. Resources are deemed strategic if they can deliver competitive advantage for the firm. Hamel and Prahalad (1994) talk in terms of core competencies that are the skills, knowledge and technologies that an organisation possesses on which its success depends. A summary of a number of these viewpoints in relation to resources is illustrated in Table 5.1.

The proliferation of many different terms to explain similar concepts can create confusion. For example, it could be argued that as well as being a capability,

Table 5.1 *An overview of terminology associated with the resource-based view*

Authors	Terminology	Description of terms
Wernerfelt (1984)	Resources	Resource position barriers
Itami (1987)	Invisible assets	Information-based resources
Dierickx and Cool (1989)	Strategic assets	Accumulation of stocks through investment
Amit and Schoemaker (1993)	Resources	Stocks of available factors owned or accessible by the firm
	Capability	The ability of the firm to deploy resources
	Strategic assets	Difficult to trade and imitate, scarce, appropriable and specialised resources and capabilities
Hamel and Prahalad (1994)	Core competence	Collective learning
		Co-ordination of production skills and integration of multiple streams of technologies
Hall (1994)	Intangible resources	Skills or competencies – including the know-how of people
		Intangible resources linked with a functional, cultural, positional, or regulatory capability
Barney (2002)	Firm resources	Assets, capabilities, competencies, organisational processes, firm attributes, information, knowledge etc. controlled by the firm

marketing know-how and experience is also a resource that the organisation possesses. In the resource-based view literature, the terms resources, activities and capabilities are also often used interchangeably. However, in the context of the outsourcing framework in this book, there is a clear distinction between resources, activities and capabilities and relationship between them. In the context of the outsourcing framework *resources* are what organisations deploy to perform *activities*. Activities refer to the routines and processes that have to be co-ordinated and integrated that enable an organisation to create and deliver products and services to their customers. These activities include the deployment of tangible and intangible resources. The term *capability* refers to the ability of the organisation to deploy resources in order to perform activities in relation to both competitors and suppliers. For example, an organisation may possess a superior cost position relative to its competitors (*a capability*) in manufacturing (*an activity*) which involves deploying equipment, people, technology etc. (*resources*). This analogy is illustrated on the hierarchy in Figure 5.3. At the bottom of the hierarchy are the resources. The resources are the building blocks of the activities necessary to deliver

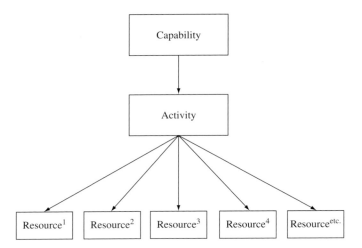

Figure 5.3 The relationship between capabilities, activities and resources

a product or service to the customer. Organisations must ensure that resources are deployed in a way that delivers both value for the customer and profit for the organisation.

Value chain analysis is a useful approach for identifying resources, activities and analysing organisation capability. Many products or services are created through a series of vertical business activities including raw material acquisition, sub-assembly manufacture, final manufacture, distribution, sales and after-sales service. Organisations deploy various resources in order to carry out business activities at each stage in the value chain. Referring back to the personal computer industry chain in Figure 5.1, it can be seen that the sale of personal computers involves a number of vertical activities including raw material acquisition, component manufacture, sub-assembly manufacture, final assembly, distribution and retailing. These activities are often referred to as an industry value chain. Each stage in the value chain deploys a number of resources in order to carry out business activities in this value chain. For example, participating in the retailing stage of the value chain involves securing access to financial resources, physical resources and human resources. There are a number of frameworks that can be used to analyse an organisation as a collection of activities depending upon the type of organisation.

5.3   The value chain

The value chain represents an organisation as a chain of activities for transforming inputs into outputs that customers value (Porter, 1985). The process of transforming these inputs into outputs involves a number of primary and support activities. Primary activities are directly involved in creating and adding value for the customer. The primary activities move from left to right in the value chain representing the activities of physically creating the product or service and transferring it to the customer. Porter (1985) has identified five primary activities associated with the value chain including the following.

- *Inbound logistics* – activities relating to receiving, storing and distributing the inputs to the product or service. They may include warehousing, inventory management and internal transportation mechanisms.
- *Operations* – activities relating to the transformation of inputs into finished products and services. Operations may include production, assembly, packaging, equipment maintenance, facilities, operations, quality assurance and environmental protection.
- *Outbound logistics* – activities relating to the distribution of finished products and services to customers. Outbound logistics may include finished goods warehousing, order processing, order picking, shipping, delivery operations.
- *Marketing and sales* – activities related to sales force efforts, advertising and promotion, market research and planning, and distributor support.
- *Service* – activities associated with providing service to enhance or maintain the value of the product. Service may include providing customer services such as installation, spare parts delivery, maintenance and repair, technical assistance, managing customer enquiries and complaints.

The primary activities are linked to a number of support activities that include the following.

- *Procurement* – refers to the activities performed in the purchasing of inputs that are used in the value chain. Procurement may take place within defined policies or procedures and involve a number of functional areas. Manufacturing and engineering have to be involved in purchasing in order to ensure the specifications and quality standards are acceptable.
- *Technology development* – includes product and process research and development, equipment design, computer software development, computer-aided design and engineering.
- *Human resource management* – involves all the activities relating to the recruitment, training, development and rewarding people throughout the organisation. Human resource management ensures that the organisation has the appropriately skilled people to carry out the activities of the value chain effectively.
- *Firm infrastructure* – includes the organisational structure, planning, financial controls, and culture designed to support the value chain.

Support activities support and improve the performance of the primary activities. Each of the primary and support activities incurs costs and should add value to the product or service in excess of these costs. Due to the sequential nature of the value chain model it is most appropriate for application in a manufacturing context where

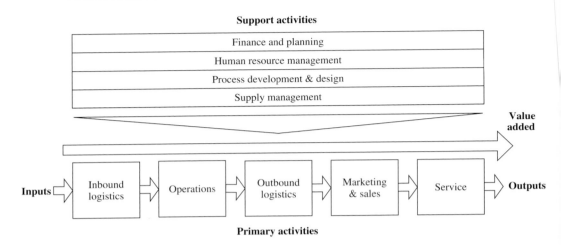

Figure 5.4 The value chain for a manufacturing-type organisation

inputs are processed into outputs right from new product development to after-sales service. Figure 5.4 illustrares the value chain of a manufacturing-type organisation. In order to deliver a low-cost or differentiated product a company must perform a series of activities. The value chain of an organisation is also a part of a value system. For example, suppliers have value chains that manufacture and deliver the inputs to the organisation's value chain. Many products pass through the value chains of channels on their way to the customer. Specialisation occurs in the value system in terms of role and capability. For example, Federal Express specialises in the distribution portion of the value system for many industries. Gaining and sustaining competitive advantage depends upon understanding both the organisation's value chain and how the organisation fits within the overall value system.

The value chain can be applied in the context of service organisations including retailers, fast-food restaurants and hotels. Haberberg and Rieple (2001) describe these as 'service-manufacturers' where a standard product is provided along with high levels of service. In these types of environments outputs are measured in volume terms while unit cost is an important measure of efficiency. The value chain is more concerned with efficiency rather than innovation-related activities such as new product development. The focus is on the process of creating and delivering the product rather than the product itself. The value chain outlines the activities that have to be performed once a product has been designed rather than developing a series of innovations.

5.4   The value shop

The nature of analysis in the value chain is very much focused on a product or service manufacturer. However, such a framework can be quite difficult to apply in the context of an organisation that provides a highly customised product or service to its customers. For example, it would be inappropriate to attempt to relate the *Inbound* or *Outbound logistics* activities to the activities of a management consultancy. Stabell and Fjeldstad (1998) have proposed the value shop framework that reflects the nature of professional service type organisations. The provision of a service has an entirely different value creation logic than Porter's value chain model. For example, service producers customise their service to the specific needs of a client rather than mass-produce a service. The value shop model is applicable in the context of professional service such as medicine, engineering, law, architecture and other professions. These organisations are labour intensive with the professionals and specialists in the problem area being frequently the largest component of the workforce. A generic value shop is illustrated in Figure 5.5.

Professional service organisations are established to solve problems for clients and provide a highly customised service. In contrast to service manufacturers, professional service organisations work closely with their clients to develop customised solutions. In many instances, professional advice and knowledge has traditionally been perceived as more important than revenue-earning capability. A professional service organisation sells the talents, skills and abilities or its professional staff. Such a service involves a high level of customisation and face-to-face interaction with the client. Professional service organisations add value by solving the problems of clients in a creative and effective manner. The value shop

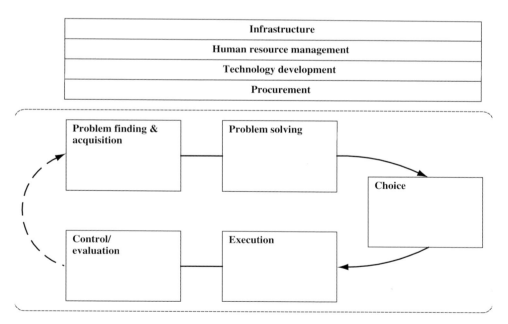

Figure 5.5 The value shop (Stabell and Fjeldstad, 1998). © John Wiley. Reproduced with permission.

schedules the activities and applies resources in a fashion that is tailored to the needs and situation of the client. As illustrated in Figure 5.5, the primary activities are performed in a cyclical fashion where the evaluation activities can be a problem-finding activity of a new problem-solving cycle. Stabell and Fjelstad (1998) have identified five primary activities associated with the value shop.

- *Problem finding and acquisition* – activities associated with the recording, reviewing, and formulating of the problem to be solved and selecting the approach to solving the problem. For example, for a general practitioner this can involve carrying out activities such as reviewing patient history, patient examination and diagnosis. This activity closely embodies marketing. For example, problem finding and acquisition for consultancies may involve a number of marketing activities such as advertising and making presentations to potential clients. Furthermore, the reputation of professionals can be a significant marketing tool.
- *Problem solving* – activities associated with generating and evaluating potential solutions. For example, a civil engineer may generate a number of designs to meet the functionality requirements for a client.
- *Choice* – activities associated with choosing among the potential solutions. Although these activities may not consume a lot of time and resource they are particularly important in terms of value. There is also considerable interaction among the functional specialists involved in the problem-solving process.
- *Execution* – activities associated with communicating, organising and implementing the chosen solution.
- *Control and evaluation* – activities associated with measuring and evaluating how implementing the chosen solution has solved the problem. For example, this may include activities such as evaluating the success of a new installed information system or safety testing in a construction project.

The support activities of the value shop are similar to the value chain as illustrated in Figure 5.5. However, it must be emphasised that many of the primary and support activities are performed together. For example, the people who obtain work for the organisation are the same people who are involved in carrying out the problem-solving and implementation processes. These same people will be involved in human resource management activities such as recruiting, developing and training professionals. The marketing, procurement, and technology development activities are also inter-linked and performed when the professionals are undertaking the problem-solving and implementation processes for their clients. Figure 5.6 illustrates an example of the value shop for a typical software house.

The value shop illustrates that the focus is on determining what the client wants and determining a means of fulfilling these needs. However, the total business does not have to be a pure service provider for the application of the value shop. Although the major focus for the manufacturer tends to be on manufacturing

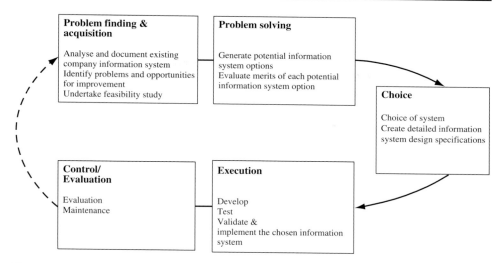

Figure 5.6 The value shop of a software house (adapted from Stabell and Fjeldstad, 1998)

the product at the lowest cost in its industry sector, in many cases a portion of the manufacturer's value chain performs similar activities. For example, many large manufacturers have new product development organisations. Even though the manufacturing part of the organisation is best analysed through the value chain, the new product development area is closely related to the value shop model.

5.5   The value network

Value networks rely on a mediating technology to link customers or users that are inter-dependent (Stabell and Fjeldstad, 1998). The value network represents a broker bringing together buyers and sellers and generating revenue and profitability by doing so. Examples of value network-type organisations are communications companies, retail banks and insurance companies. For example, a retail bank links customers who have money to invest with customers who want to borrow money. The term value network emphasises that a critical determinant of value to any customer or user is the network of customers or users that are connected to it. An interesting illustration of a value network is a virtual community on the Internet. A virtual community allows individuals with common interests and opinions to communicate freely and post information on relevant issues. Rheingold (1993) has defined a virtual community as social aggregations of a critical mass of people on the Internet who engage in public discussions, interactions in chat rooms, and information exchanges with sufficient human feelings on matters of common interest to form webs of personal relationships. Virtual communities usually have chat rooms, message boards and discussion groups for extensive member interaction and

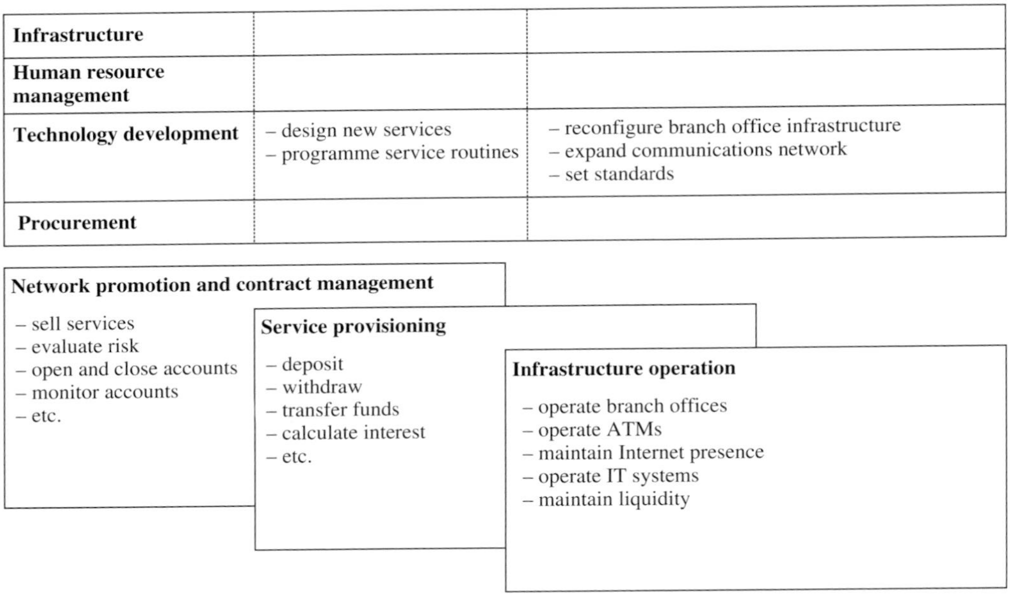

Infrastructure		
Human resource management		
Technology development	– design new services – programme service routines	– reconfigure branch office infrastructure – expand communications network – set standards
Procurement		

Network promotion and contract management

– sell services
– evaluate risk
– open and close accounts
– monitor accounts
– etc.

Service provisioning

– deposit
– withdraw
– transfer funds
– calculate interest
– etc.

Infrastructure operation

– operate branch offices
– operate ATMs
– maintain Internet presence
– operate IT systems
– maintain liquidity

Figure 5.7 The value network of a retail bank (adapted from Stabell and Fjeldstad, 1998)

they are characterised by a significant amount of user-generated content. For example, BioMedNet, a virtual community of biologists and the medical community, features specialist discussion groups, a regular forum on a topical issue, a research database called Medline, a job exchange and a daily newsletter. In order to increase value for the potential user these types of arrangements must achieve scale and capacity. The same analogy applies for a newspaper that connects journalists with readers and advertisers with prospective customers. The more readers the newspaper gains then the more advertisers it will be able to attract. In effect, the intermediary generates revenue and profit if both buyers and sellers perceive value from that intermediary. In effect, customers must perceive that there is value in being 'connected' to the intermediary. The value network for a retail bank is shown in Figure 5.7.

The primary activities are listed here.

- *Network promotion and contract management* – involves carrying out the following activities.
  - *Marketing and promoting the network to potential customers or users* – for a special interest web site (such as Soccernet.com or Teamtalk.com), this would involve advertising with complementary interest web sites or charging fees for specialist information. For a retail bank, this would involve attracting and evaluating customers or organisations and encouraging them to borrow money from the bank.

- *Marketing and promoting the network to potential providers* – for a special interest web site, this would involve selling advertising. For a retail bank, this would involve persuading individuals or organisations to deposit money.
- *Selection of customers or users* – for example, checking the credit history of a potential customer.
- *Management and termination of contracts* – for example, opening and closing customer accounts or subscriptions.
- *Service provisioning* – activities associated with establishing and maintaining links between customers and billing for value received. These activities focus on the operation of the network on a daily basis encompassing activities such as billing customers and handling customer queries and complaints. In the context of a special interest web site activities would include the operation of the servers and communication mechanisms between the users and providers.
- *Network infrastructure operation* – activities associated with maintaining and running the physical infrastructure. The network infrastructure operation activities depend upon the type of organisation. In a mobile telephone company the key infrastructure is the network, while in a retail bank it is the branch offices, computing systems and financial assets. In effect, it is concerned with being in a state of alertness to deal with a range of customer requirements. For a retail bank or building society it includes operation of branch offices, automatic teller machines, a fully functional web site, links with other service providers (such as insurance and legal and surveying services etc.), and adequate funds on-hand to meet the needs of customers.

There may be a certain amount of overlap between these three primary activities. For example, the rapid evaluation of a prospective customer (*Network promotion and management activity*) can enhance the level of service (*Service provisioning activity*) that the organisation is providing. Furthermore, as shown in Figure 5.7, there is overlap between the primary activities emphasising the concurrent nature of the inter-relationships. For example, in the telecommunications industry network operators provide the infrastructure for service providers who also act as communication infrastructure for payment and billing services. The support activities are similar to those outlined in the value chain for a manufacturing organisation. There are important linkages between the primary activities. The *Technology development* support activity may involve creating innovations directly impacting the infrastructure operation primary activity. For example, by designing and establishing an innovative Internet presence for customers, a building society is contributing to the further development of its infrastructure operation. In addition by establishing an Internet presence this may enable customers or organisations to transfer funds electronically in a virtual environment. Thus, the service provisioning activity is being directly enhanced.

Illustration 5.2

Network economics at eBay

The key characteristic of the value network model is the presence of network economics. Network economics, as expressed by Metcalfe's Law, describes a situation in which the value of a product or service increases as a function of how many other users are using the product or service. Bob Metcalfe, the inventor of the Ethernet, proposed that the value of a network to each of its users is proportional to the number of other users (which can be expressed as $(n^2-n)/2$). In effect, the value to the customer is determined by the number of other individuals who adopt the technology. In recent years, the presence of network economics has been considered to be the primary determinant of competitive advantage in high technology industries such as computer hardware and software, telecommunications, consumer electronics and Internet services. Classic examples of network economics include the telephone and fax machine. Network economics contrasts with the traditional diminishing returns view of how markets and businesses operate as proposed by many economists. The assumption of diminishing returns is that products or companies that achieve a position of advantage in a market eventually encounter problems so that a predictable equilibrium of prices and market shares is reached. This assumption is contrary to the key characteristic of network economics – positive feedback. The Internet best illustrates the effects of the presence of positive feedback. The value of connecting to the Internet for users is a function of the supply of useful information that can be accessed and the opportunity for carrying out commercial transactions over the Internet. Moreover, the number of individuals connected to the Internet amplifies the supply of information and commercial transactions.

One company that has successfully exploited the network economics effects and positive feedback of the Internet has been eBay.com. eBay is the world's largest online auction service with millions of unique auctions and over 500 000 new auction items being added daily. eBay provides the electronic infrastructure to manage the auctioning process between millions of buyers and sellers. The company derives its revenue from collecting an upfront submission fee, plus a commission as a percentage of the sale value. The amount of the submission fee is determined by the amount of exposure that the seller requires for their item. For example, featuring the item in a specific product category or on the eBay homepage requires the payment of a higher fee. When a successful bid has been made, the buyer and seller negotiate the payment method, delivery

Table 5.2 *Value network elements of eBay*

Value network components	
Network promotion and contract management	• Liquidity – a critical mass of buyers and sellers • Brand • Reputation • Referrals
Service provisioning	• Wide ranges of products and services • Reliability • Security • Guarantee
Network infrastructure operation	• Regional sites • User interface • Technology infrastructure

details, and warranty. eBay acts as the interface between the buyers and sellers and has no involvement in the management of physical inventory or delivery management. Through its seller payment protection scheme sellers are protected against fraudulent credit card purchases through services such as secure processing, credit card charge-back protection and privacy protection. eBay has also added a number of sites to its portfolio including eBay Motors and eBay Sports and eight regional sites to limit the difficulties associated with delivering bulky items. eBay derives the majority of its revenue from the fees and commissions associated with the auctioning process. However, one of eBay's subsidiaries, Half.com, offers a fixed-price, buyer-to-seller facility across a range of products including CDs, books, DVDs, and games.

eBay has exploited the network economics characteristics nature of the Internet to build a significant competitive advantage that has been extremely difficult for other companies to replicate. As one of the first movers in the online auction business, it developed a network of loyal buyers and sellers. Even though other companies such as Amazon and Yahoo! have followed the lead of eBay by offering auction services, they have found it difficult to erode eBay's competitive position. By being one of the first firms to enter this business, eBay built up significant liquidity in the form of a critical mass of buyers and sellers. Buyers use the site because they know they will find the broadest range of products, whilst sellers use the site as they know they will obtain the highest price due to the high number of buyers. Table 5.2 relates eBay to the value network model and outlines the various elements that have enabled it to build a strong competitive position.

eBay is an example of a company that can be related entirely to the value creating logic of the value network model. In contrast, Amazon's business has elements that relate to both the value chain model (logistics management) and the value network model (buyer-seller fixed-price matching). eBay has been able to build and sustain its competitive position by widening its range of products and services and establishing alliances with providers of complementary services. Indeed, the relative ease with which alliances can be formed by companies on the Internet for the provision of products and services contrasts with that of the bricks-and-mortar environment. The development of eBay illustrates how the characteristics of the positive feedback loop can give a first mover a significant advantage with other competitors being marginalised. Furthermore, the presence of network economics effects can create significant barriers to entry into the market-place due to the dominant position of the leading company.

5.6   Business process perspective

At this stage it is worth relating the business process approach to mapping the organisation with that of the activity approach. Business processes are a set of activities that cut across an organisation, encompassing all the traditional business functions from marketing, operations, finance, procurement and human resource management as illustrated in Figure 5.8.

Business processes are not the responsibility of any one function but can involve a number of functions. Each business process normally includes a number of sub-processes, each of which involves a number of business functions as illustrated in the following examples.

- *Customer service* may involve sales enquiry handling, order processing, order fulfilment, delivery, after-sales service, etc.
- *New product development* may involve market research, analysis of competitor offerings, process design, detailed design, engineering approval, prototype testing, etc.
- *Supply chain management* may involve inbound and outbound logistics management, cost analysis, supplier relationship management, supplier development, early supplier involvement in new product development, etc.

The business process approach is normally considered along with re-engineering. Re-engineering is the fundamental rethinking and radical redesign of business processes to achieve dramatic improvements in critical measures of performance such as cost, quality, service and speed (Hammer and Champy, 1993). Organisations should completely rethink their business and focus on a collection of business processes rather than a set of business functions.

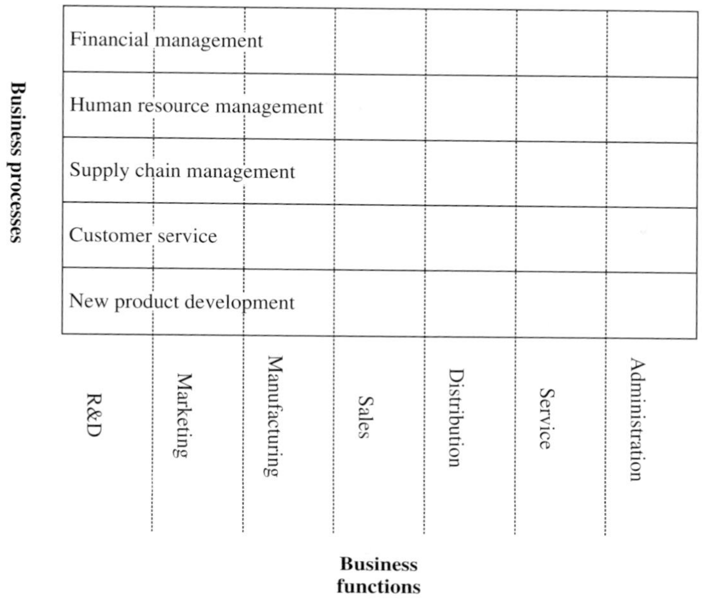

Figure 5.8 Business processes

There is considerable overlap and commonality between the activity and business process perspectives. If one refers to the sub-activities within the activities of the value chain, it can be seen that in some cases these are comparable to processes. For example, the order processing sub-activity within the outbound logistics activity and the assembly process sub-activity in the operations activity are comparable to processes. Although there are differences in terminology between the activity and business process perspectives, they are being employed to describe similar phenomena. For purposes of analysis in this book the terminology and approach adopted will refer to activities and sub-activities. As will be seen in the *Capability Analysis* in *Chapter* 7, the important point is to identify activities or sub-activities that enable both internal and external comparisons to be made on performance. Activity analysis facilitates a structured benchmarking approach to assessing an organisation's capabilities in the range of activities identified by management in relation to the potential suppliers and competitors' capacity to provide these activities. In effect, each company is in competition with all potential suppliers of each activity in its value chain. In the context of the outsourcing decision, knowing the capabilities of your competitors can be as important as knowing their relative market shares. Each point in the value chain relevant to providing the activity should be analysed. Therefore, each selected activity will be benchmarked against all potential external providers of that activity. The basic nature of strategic analysis may move from an industry analysis to a horizontal analysis of capabilities across all potential providers of the activity regardless of which industry the provider might be in. For example, these activities may include logistics, warehousing, process design, product design, quality control and plant engineering as well as the actual production that may be a function of the workforce skills.

5.7 Depth of analysis

It is important to clarify the level of detail that activity analysis in outsourcing evaluation can take. A common analytical problem is to delve into too much detail and produce a lot of information that cannot be understood. The purpose of segmenting the organisation into a range of activities is therefore not so much to simplify the analysis as to simplify the understanding that follows. Segmentation of value chains into activities and sub-activities is valuable from a number of perspectives:

- Performance in the activity has a significant impact upon whether the customer of the organisation buys their products or services. For example, many organisations are investing considerable resources in new product development as it can create the potential for the development of a stream of innovative products.
- The activity represents a significant or growing proportion of cost of producing the product or delivering the service. For example, sub-assembly activities (part of the operations activity) in manufacturing can be the largest single cost associated with manufacturing the product.
- Segmenting an activity into a number of sub-activities may allow an organisation to identify opportunities for outsourcing parts of the activity rather than the entire activity. For example, an activity such as after-sales service is composed of a number of sub-activities including on-line product support, field service support, and spare parts ordering, etc. Analysis at this level can enable the identification of potential suppliers to provide some of these services.

Categorising activities requires judgement with there being no hard and fast rules. If it is thought provoking, then the understanding of the boundary of the organisation is probably being enriched. The depth of evaluation of the organisation's value chain can take place at the activity or sub-activity level depending upon the particular circumstances of the organisation. In the case of a manufacturing company the activity level would be for example, Inbound logistics, while a sub-activity (within Inbound logistics) would be Material handling. This is illustrated in Figure 5.9.

In the case of a manufacturing company, the outsourcing decision may refer to closing down part of the business – for example, the plastic injection moulding shop – and outsourcing the supply of the relevant items to external suppliers. In certain circumstances it is necessary to consider outsourcing practice at a more

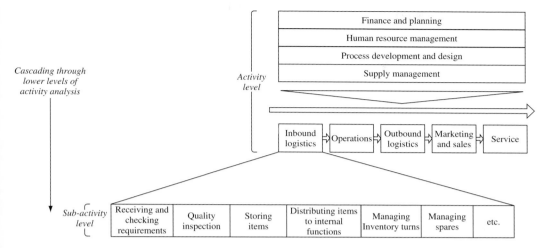

Figure 5.9 The levels of evaluating relevant organisational activities

detailed level in the value chain. For example, an activity increasingly being outsourced by companies is the recruitment of key people to specialist recruitment agencies. This example is at a much more detailed level of analysis in the outsourcing decision in contrast to the previous example. However, analysis from an activity perspective in each case is critical given that companies are increasingly positioning themselves in specific parts of the industry chain in order to gain competitive advantage. Therefore, companies considering outsourcing must rigorously evaluate their company's capabilities in relation to both their suppliers and competitors.

As well as segmenting the organisation into a number of key activities and sub-activities it is necessary to understand the linkages that exist between activities. The relationship between the organisation and its customers and suppliers should not be perceived as a series of independent activities. Behaviour in one part of the organisation can affect the costs and performance of other business functions. For example, the acquisition of more expensive raw materials and greater emphasis on quality inspection will increase inbound logistics and operations costs; however, such costs may lead to greater savings in after sales service activity.

A number of activities and sub-activities depend upon each other. For example, the co-ordination and integration of inbound logistics, operations and outbound logistics can be a source of competitive advantage through lower costs and a more responsive service. There are also linkages between the primary and support activities. For example, procurement directly affects the performance of the inbound logistics activity whilst product and process design have a direct impact upon the operations activity. In some cases, these linkages can be a source of competitive advantage. Applications of information technology have enabled organisations to link disparate activities to provide higher levels of service for customers. So far, the discussion has centred on the linkages that exist internally. However, external linkages between the organisation's customers and suppliers should be identified. For example, on-line seat reservation systems for 'no-frills' airlines such as Easyjet and Ryanair have facilitated the reduction of costs and the speeding up of the reservation activity for customers. Innovations in Internet technologies such as 'intranets' and 'extranets' are critical in integrating and co-ordinating cross-functional teams across organisational boundaries. For example, at the buyer–supplier interface, organisations have embraced Internet technologies to link their order processing systems with their suppliers. Information technologies can also eliminate activities or sub-activities that have traditionally being carried out manually. For example, in the procurement activity, less resource and time has to be expended in the processing and expediting of orders.

The mapping of an organisation to a particular configuration and identifying its relationship with its environment is an important starting point in outsourcing evaluation. Adopting an activity perspective can assist in analysing outsourcing at the following levels.

- *Perform the entire activity internally* – this involves the sourcing organisation continuing to perform the activity internally. This may involve in investing and developing the activity in order to maintain and enhance any performance superiority possessed by the sourcing organisation. This is sometimes referred to as *insourcing*.

- *Outsource the entire activity* – this involves the sourcing organisation outsourcing the entire activity to an external supplier. This may involve staff and equipment being transferred from the sourcing organisation to the supplier. The supplier is responsible for performing the activity and delivering the associated product or service to the sourcing organisation. In this case, the chosen supplier may possess the capabilities to undertake the entire activity in-house. Alternatively, part of the outsourcing agreement will involve the supplier managing and co-ordinating a number of suppliers in order to deliver the product or service. For example, the trend towards modularisation in manufacturing has led many original equipment manufacturers to outsource sub-assembly activities and source them from a single supplier who becomes responsible for co-ordinating the activities of suppliers at lower levels in the supply chain.

- *Partial outsource* – this involves the sourcing organisation outsourcing a number of sub-activities that comprise the entire activity. In this case, the sourcing organisation has identified a number of competent suppliers to perform the relevant sub-activities. Pursuing this approach, the sourcing organisation avoids being over dependent upon a single supplier. The sourcing organisation may also establish a relationship with each of the sub-suppliers in order to foster competition and obtain lower prices.

Table 5.3 *Characteristics of value configurations (adapted from Stabell and Fjeldstad, 1998)*

	Value chain	Value shop	Value network
Value creation logic	Processing of inputs into outputs	Solving customer problems	Linking customers
Primary technology	Long-linked	Intensive	Mediating
Primary activities	• Inbound logistics • Operations • Outbound logistics • Marketing • Service	• Problem finding and acquisition • Problem solving • Choice • Execution • Control/evaluation	• Network promotion and contract management • Service provisioning • Infrastructure operation
Interactivity relationship logic	Sequential	Cyclical	Simultaneous
Primary activity interdependence	• Pooled • Sequential	• Pooled • Sequential • Reciprocal	• Pooled • Reciprocal
Example organisations	Manufacturers, retailers, fast-food outlets, hotels, etc.	Software house, architects, engineering, medicine, consultancies, etc.	Retail bank, insurance company, postal services, etc.
Key cost drivers	• Scale • Capacity utilisation		• Scale • Capacity utilisation
Key value drivers		• Reputation	
Business value system structure	• Inter-linked chains	• Referred shops	• Layered and inter-connected networks

A range of supply relationships – ranging from adversarial to close collaborative – can be established in order to pursue either of the outsourcing options. Primarily, the importance of the activity being outsourced and the level of risk in the supply market will influence the supply relationship chosen. The appropriate supply relationships are discussed in more detail in *Chapter 9 – Developing the relationship strategy*.

Segmenting the organisation into activities also provides a framework for analysing the shifting responsibilities for roles in the value chains of both customers and suppliers. As will be seen in the following chapters, activity analysis provides a framework for evaluation of specialisation and focus in particular activities. It creates an awareness of the impact of the performance of the key activities identified on business success. For example, an organisation may discover from this analysis that it has over-committed in certain areas because it is performing too many activities internally. This may create a situation where the organisation is under-performing in a number of activities that are critical to business success. This analysis is relatively straightforward to understand and relate to most organisations. For example, the three models have illustrated the ease with which the models can be adapted and applied to reflect the requirements and situation of an organisation. The categorisation of activities outlined in each model is for guidance only. Table 5.3 provides a summary of the characteristics of each model. The categories in each model do not have to be rigidly adhered to and can be adapted to suit the requirements of the relevant organisation. Indeed, as will be shown in the following chapter, in outsourcing evaluation and management it is more important to determine the importance level of activities rather than segmenting the organisation into primary and support activities.

REFERENCES

Alexander, M. (1997). Managing the boundaries of the organisation. *Long Range Planning*, **30**, No. 5, 787–789.

Amit, R. and Schoemaker, P. J. H. (1993). Strategic assets and organisational rent. *Strategic Management Journal*, **14**, No. 1, 33–45.

Barney, J. A. (2002). *Gaining and Sustaining Competitive Advantage*. New Jersey: Prentice Hall, second edition.

Collins, R. and Bechler, K. (1999). Outsourcing in the chemical and automotive industries: choice or competitive imperative? *Journal of Supply Chain Management*, **35**(4), 6 (Fig. 2).

Dierickx, I. and Cool, K. (1989). Asset stock accumulation and sustainability of competitive advantage. *Management Science*, **35**, No. 2, 1504–1511.

Haberberg, A. and Rieple, A. (2001). *The Strategic Management of Organisations*. London: Financial Times – Prentice Hall.

Hall, R. (1994). A framework for identifying the intangible sources of sustainable competitive advantage. In: *Competence-based Competition*, Hamel, G. and Heene, A.(eds.). Chichester: Wiley, pp. 149–69.

Hamel, G., and Prahalad, C. K. (1994). *Competing for the Future*. Boston: Harvard Business Press.

Hammer, M. and Champy, J. (1993). *Re-engineering the Corporation*. New York: Harper-Collins.

Itami, H. (1987). *Mobilising Invisible Assets*. Cambridge, MA: Harvard University Press.

Porter, M. E. (1985). *Competitive Advantage*. New York: Free Press.

Rheingold, H. (1993). *The Virtual Community: Homesteading on the Electronic Frontier*. Reading, MA: Addison-Wesley.

Stabell, C. B and Fjeldstad, O. D. (1998). Configuring value for competitive advantage: on chains, shops, and networks. *Strategic Management Journal*, **19**, 413–37.

Wernerfelt, B. (1984). A resource-based view of the firm. *Strategic Management Journal*, **5**, No. 2, 171–180.

6 Activity importance analysis

6.1 Introduction

This stage in outsourcing evaluation involves determining the level of importance of the activities that have to be performed in order to satisfy customer needs. It is crucial for an organisation to determine which activities are critical to competitive advantage. Identifying critical activities involves understanding the major determinants of competitive advantage in the markets or the industries in which the organisation competes or might wish to compete. An analysis of the competitive environment and customer needs can have an important role in identifying which activities are critical for success in the business environment. Critical activities will enable an organisation to differentiate itself from its competitors in the way in which it serves its customers. An understanding of critical activities is central to outsourcing evaluation. For example, if an organisation possesses a superior capability in a critical activity relative to competitors or suppliers, then it should continue to perform that activity internally. The organisation must also have a clear understanding of how sustainable this position is over time. Alternatively, where possible, activities that are not key influences on the ability of the organisation to achieve competitive advantage should be outsourced. Critical activities are those that can be used to build sources of advantage that are difficult and costly for competitors to replicate. Focusing attention on customer needs and competitive advantage will involve the organisation applying its distinctive capabilities to meet the needs identified. For example, in the beer market where branding is crucial to competitive advantage, companies such as Diageo place great emphasis on their advertising and promotional strategies in order to give their products a strong brand image. Maintaining control of critical activities – either through ownership or collaborative relationships – will enable the organisation to maintain and enhance its competitive position. This process can be carried out on a formal or on-going basis in response to changing circumstances.

6.2   Background to the importance level analysis

This stage of the analysis is influenced by elements of the industry view. This section outlines the approach taken in the outsourcing framework in the context of some of the key elements of the industry view. In particular, some of the key similarities and differences are identified in relation to Porter's generic strategies and the definition of critical activities.

6.2.1   The generic strategies

Michael Porter (1985) has argued that there are three generic strategies that organisations can pursue in order to obtain competitive advantage and outperform competitors. These strategies are given here.

- *Cost leadership* – this involves the organisation being the lowest cost producer of products or services in an industry. The organisation can charge a lower price than competitors and be in a stronger position to defend its position if competitive rivalry increases. The success of such a strategy depends primarily upon having strong capabilities in cost reduction across all aspects of the business.
- *Differentiation* – this involves producing a product or service in a way that is perceived by customers as offering a unique value proposition. Organisations that pursue this strategy often have lower market shares and are normally targeting customers that are not price sensitive. Successful differentiation strategies are based on possessing skills in areas such as research and design, product development and marketing.
- *Focus* – this strategy involves serving the needs of a particular market segment. The organisation concentrates on a market niche based upon either the needs of a unique segment of customers or a narrowly defined geographic market. Once a market segment is selected, the organisation pursues either a cost leadership or differentiation strategy.

Porter argues that attempts at combining any of these strategies will lead to failure and an organisation being 'stuck-in-the-middle'. These recommendations have become a significant cornerstone of strategic thinking. The generic strategies capture a great deal of the complexities associated with strategic choice. The cost leadership or differentiation strategies can be used to describe a complex range of the key elements of business strategy. The generic strategies provide a valuable and straightforward means for describing the complexities associated with business strategies. However, this is one of the critical flaws of the generic strategies. It is over simplistic to place the overall strategy of the organisation into narrow categories. Furthermore, Porter's hypothesis that organisations that are 'stuck-in-the-middle' are unprofitable has attracted considerable criticism. Although research has confirmed that the strategies of cost leadership and differentiation advantage can adequately describe the strategy of an organisation, there is evidence to show that successful organisations can pursue a combination of the cost leadership and differentiation strategies. In fact, Miller and Dess (1993) have found that companies that pursue a combination of the generic strategies can be profitable. It is possible to pursue a differentiation strategy that can lead to an organisation being the lowest cost producer in the industry. A differentiation strategy will allow the organisation to offer customers a superior product or service which in turn leads to high market share and the associated benefits of economies of scale and experience. The Japanese car companies such as Toyota or Honda are often used as exemplars of successfully integrating both generic strategies. Organisations can design quality into their products in a way that leads to a combination of higher quality, higher market share and lower costs (Kroll *et al.* 1999). Furthermore, the resource-based theorists argue that competitive advantage is based upon building capabilities that facilitate the development of innovative products and services rather than positioning the company in order to pursue cost leadership or differentiation.

The approach proposed in the outsourcing framework presented in this book is that organisations can pursue a combination of the generic strategies. Evaluating outsourcing from an activity perspective supports this approach. In many cases, the nature of the activity being outsourced will determine whether the sourcing organisation is impacting the cost or differentiation dimensions. Sourcing decisions related to administrative functions or standard manufacturing processes with little value added will impact upon the cost structure of the organisation. Alternatively, sourcing decisions related to activities with a significant creative element such as design are likely to impact the differentiation dimension because they have the potential to lead to innovations that allow the organisation to differentiate its offerings in the eyes of the customer. However, regardless of whether the cost or differentiation dimension is being impacted, outsourcing evaluation and management must be based upon supporting the overall strategy of the organisation. Although outsourcing considerations can be part of the strategy formulation process, it is more likely that outsourcing will be guided by the overall strategy of the organisation that has already been formulated. However, there is considerable overlap between the analysis carried out in strategy formulation and outsourcing evaluation. For example, understanding the key determinants that influence customers on whether they buy the products or services of the organisation is also a major element of any business strategy.

6.2.2 Distinguishing between critical and non-critical activities in the context of outsourcing

The frameworks for value chain analysis presented in Chapter 5 segment an organisation into primary and support activities. However, in the context of outsourcing it is also important to determine the importance level of activities rather than segmenting the organisation into primary and support activities. For outsourcing purposes, critical and non-critical activities are defined as follows.

- *Critical activities* – have a major impact upon the ability of an organisation to achieve competitive advantage either through the ability to reduce cost and/or create differentiation. Therefore, superior performance in such an activity relative to competitors offers customers a unique value proposition. Consequently, the activity is a source of competitive advantage. For example, an engine manufacturer may compete in markets where customers are increasingly demanding high standards of fuel economy, emissions and engine performance. Therefore, fuel systems and pistons are regarded as critical components – activities that are critical to fuel economy, emissions and performance. Critical activities can reside entirely within the organisation. However, a critical activity can also extend across organisational boundaries. For example, an organisation may work jointly with one of its key suppliers in order to develop a critical process that increases the functionality and reduces the cost of its product portfolio.

- *Non-critical activities* – have a limited impact upon the ability of an organisation to achieve competitive advantage. Although these activities are central to successfully serving the needs of customers in each market, any performance advantage obtained in such activities will not lead to a sustainable competitive advantage as competitors or suppliers can easily replicate this performance advantage.

Distinguishing between critical and non-critical activities in the context of outsourcing is important for the following reasons.

- Assessing the importance level of activities allows the organisation to determine whether outsourcing an activity will maintain the competitive position of the organisation or act as a source of competitive advantage. For example, outsourcing an activity such as catering or security that has limited or no impact upon why customers buy the products or services of the organisation is unlikely to have any impact upon competitive differentiation. Alternatively, employing outsourcing to achieve superior performance in a critical activity has the potential to create competitive advantage. For example, an organisation may outsource an activity whilst employing astute relationship mechanisms – either competitive or collaborative – that allow it to leverage potential benefits from a supplier that are unavailable to its competitors.

- The importance level of an activity is a valuable indicator of the level of resource and attention that should be given to managing that activity. This approach is quite distinct from the core/non-core logic. Using the core/non-core logic, there is a tendency for organisations to define all outsourced activities as non-core and, as a consequence, of low importance. However, employing definitions of critical and non-critical activities in this context provides a valuable basis for

determining the level of attention that should be given to managing the outsourcing process. For example, a higher level of attention should be given to an
outsourcing strategy that is designed to strengthen a critical activity than the
outsourcing of a non-critical activity.

- The importance level of activities are linked to the factors in the external
business environment that can create business success. These factors can also
change over time due to changes in customer preferences or improved competitive offerings. For example, an organisation may decide to perform an activity
internally because it is currently important in the eyes of the customer; i.e. one of
the principal reasons why the customer purchases the product. However, over
time that factor may diminish in importance in the eyes of the customer as
suppliers become more proficient at performing the activity; i.e. it is no longer a
source of differentiation. In this case, it will be more prudent to outsource the
activity. Employing this approach allows outsourcing evaluation and management to be linked with the factors that create business success.

Although the overall strategy of the organisation will provide valuable direction
and inform the outsourcing evaluation and management process, this chapter
provides a number of techniques that can assist in determining the importance
level of organisational activities. These techniques can be employed and integrated
alongside the overall business strategy adopted by an organisation. Firstly, it is
shown how Porter's Five Forces model can assist in understanding the competitive
environment. By linking an understanding of the competitive environment with
outsourcing evaluation, it is shown how it is possible to obtain an indication of the
activities – both internal and external – that are central to competitive advantage.
As well as the competitive environment, it is shown how understanding the
concept of customer value is a useful basis for determining the importance of
activities. Linking an understanding of the concept of value and the competitive
environment will enable an organisation to determine the critical determinants
that ensure success. Furthermore, the critical success factors (CSFs) methodology
is useful in an outsourcing context in that it can establish a direct link between
outsourcing evaluation and the factors that influence business success. For
example, critical success factors are a very effective tool for providing information
about customer needs and potential sources of competitive differentiation. The
identification of critical success factors can enable an organisation to determine
the key activities that create value. Such analysis can establish links between
environmental opportunities and threats and the strengths and weaknesses of
the organisation. The development of certain activities – either internally or
externally with suppliers – can enable an organisation to exploit market opportunities and in some cases create new opportunities in the competitive environment.
By employing these techniques in outsourcing evaluation it is possible to obtain
valuable insights into the types of capabilities that are likely to secure competitive

advantage. This analysis of the importance level of organisational activities serves as an introduction to the analysis of the capability of the sourcing organisation in relation to competitors and suppliers. There is also a considerable degree of iteration between these two stages. For example, the capability analysis in a critical activity may reveal that the sourcing organisation is the most competent in comparison to competitors and suppliers, which further enhances the importance of the activity. In fact, it is quite possible for an organisation to undertake the capability analysis in a certain activity before determining the importance level of the activity. However, the advantage of determining the activity importance first is that the level of importance will determine the amount of resource and time that should be allocated to determining the capability in the activity.

Illustration 6.1

An analysis of outsourcing and its effects on organisational performance

A study carried out by Matthew Gilley and Abdul Rasheed examined the influence of outsourcing on the financial and non-financial performance of organisations. The rationale for carrying out this study was that much of the literature on outsourcing and its impacts on performance was based primarily on anecdotal evidence. The distinction between peripheral and core activities was employed as a basis for the analysis. They analysed the moderating effects of both company strategy and the nature of the environment on outsourcing and performance. As well as considering these dimensions, they also developed a construct that they termed 'outsourcing intensity' to provide a clear indication of the reliance of an organisation on outsourcing. This construct was derived from two fundamental properties, *breadth* and *depth*. *Breadth* referred to the number of activities (finance, information systems, manufacturing etc.) outsourced as a percentage of the total number of activities that a company could perform. *Depth* referred to the extent to which an organisation outsources a particular activity. For example, an organisation contracting out a high proportion of the value of each outsourced activity is regarded as having a 'deeper' outsourcing strategy. The study analysed the influence of outsourcing intensity on financial (return on assets, return on sales, and overall financial performance), innovation (research and development outlays, process innovations and product innovations) and stakeholder performance (employment growth/stability, employee morale, customer relations and supplier relations).

Their initial propositions before carrying out the study were that company performance was influenced in different ways depending upon whether peripheral or core activities were being outsourced. One of their initial propositions was that peripheral outsourcing had a positive impact upon performance whereas core outsourcing had a negative impact upon performance. This was based on the predominant view in the literature that outsourcing peripheral activities leads to cost reductions and allows firms to focus on more critical business activities. Alternatively, outsourcing near-core activities leads to a loss of innovation potential and the threat of competition from suppliers. The influence of the firm's generic strategy (i.e. cost leadership or differentiation) was also considered in the context of the outsourcing of peripheral and core activities. In relation to firms pursuing a cost leadership strategy, they proposed that outsourcing would have a positive impact on their performance. Alternatively, firms pursuing a differentiation strategy would gain less by outsourcing peripheral activities, as any cost improvements obtained would have little direct impact upon differentiation. They also proposed that firms pursuing a cost leadership strategy will reduce the negative impacts of outsourcing core activities as they can access the lowest cost source for the activity. Alternatively, firms pursuing a differentiation strategy will further exacerbate the negative effects of outsourcing core activities as they are likely to lose control of potential sources of innovation that allows them to pursue their differentiation strategy.

In some instances, their initial propositions were supported by the findings, whilst in others the findings were contradictory. Indeed, their findings in relation to the impact of outsourcing on firms pursuing either a cost leadership or differentiation strategy were most interesting. In relation to organisations pursuing a cost leadership strategy, it was found that there was a positive relationship between outsourcing and performance. Outsourcing peripheral activities had a positive impact upon financial performance and outsourcing core activities had a positive impact upon the innovation performance of organisations pursuing a cost leadership strategy. These findings support the prevailing view within much of the literature that outsourcing can be employed to enhance the overall cost position of the organisation. Similar results were obtained for organisations that were pursuing a differentiation strategy. Organisations that outsourced a higher level of peripheral activities experienced higher levels of performance along the innovation dimension. Therefore, the logic of this finding is that through outsourcing peripheral activities, organisations can focus greater attention on the more innovative activities. Most interestingly, contrary to their initial proposition, they found that differentiation strategies were compatible with outsourcing core activities. This is an interesting finding as it contradicts much of the literature that argues that outsourcing can lead to a loss of competitiveness in innovation-related activities.

Source: Gilley, K. M. and Rasheed, A. (2000). Making more by doing less: an analysis of outsourcing and its effects on firm performance. *Journal of Management*, **26**, No. 4, 763–90.

6.3   Understanding the competitive environment

6.3.1   Porter's five forces analysis

Understanding the competitive environment is a useful starting point in determining the importance level of organisational activities. Outsourcing evaluation and management is influenced by the competitive conditions that are prevailing in the business environment. Michael Porter's five forces analysis is probably the most widely used and influential model for understanding the competitive environment. Porter (1980) argues that the essence of strategy formulation is coping with competition. Competition in an industry is rooted in its underlying economics and competitive forces that can extend well beyond the established competitors in a particular industry. For example, customers, suppliers, new entrants, and substitute products or services are all potential competitors with their competitive stance being significantly influenced by the industry in question. The collective strength of each of these forces determines the profit potential of an industry. The type of industry structure will impact the potential for profit. For example, in an industry in which entry is relatively straightforward, the prospects for long-term profitability are limited. Conversely, in an industry where the competitive forces are weak, there are likely to be greater opportunities for profit. The objective for a company is to determine how it can best defend itself against the five forces or how it can influence them in a way, which positively impacts its competitive position. The challenge is to analyse and understand the sources of each force. Understanding these sources of competitive pressure provides a basis for strategic action. The strongest competitive force or forces will have the greatest impact upon profit levels within the industry. For example, in the UK food-processing industry the immense buying power of the leading supermarket chains is the most significant force. In this scenario, dealing with this force becomes the major strategic priority for the food processors. Therefore, certain forces are more important in influencing competition depending upon the industry. Each of the five forces is now discussed in turn.

6.3.1.1 Threat of entry

The potential threat of entry depends upon the barriers present within the industry as well as the likely reaction from existing competitors to entry. New entrants can pose a challenge to existing competitors by creating more capacity and attempt to take market share. In an industry where entry barriers are high, existing companies will react fiercely to any new entrants in order to protect their position. In fact, the anticipated reaction of existing competitors is likely to be the most significant influence on the entry decision. Porter (1980) has identified the six major sources of barriers to entry as follows.

- *Economies of scale* – scale economies in production, research, marketing and service are key barriers to entry in many industries. For example, entry into the car industry will involve considerable resource in terms of finance and experience in order to obtain a competitive cost base. Economies of scale can also be in the form of distribution, financing and after-sales service.
- *Product differentiation* – for example, a strong brand within an industry can be a significant barrier to entry. Advertising, promotion, customer service, reputation and product uniqueness are all potential contributors to the building of a brand.
- *Capital requirements* – a huge financial requirement in order to establish a competitive position can be a significant barrier to entry in an industry. For example, in the aircraft industry there are high capital requirements. Entry to many industries is only open to large global corporations.
- *Cost disadvantages independent of size* – companies within an industry may have cost advantages that are not attainable by new entrants, regardless of their size and economics of scale. Experience and knowledge that is accumulated over time can be extremely difficult to replicate. For example, Intel's skills and experience in the design and manufacture of microprocessors is a significant barrier to entry in to the microprocessor industry.
- *Access to distribution channels* – this is a key factor for any company entering an industry. In industries where existing competitors exercise considerable control over access to distribution or retail channels, barriers to entry are likely to be high.
- *Government policy* – can also act as a barrier to entry into an industry by restricting licences, issuing exclusive franchises, or establishing legislation that is costly and cumbersome to implement. Government policy can also be in the form of protectionist legislation preventing foreign competitors establishing a foothold in local markets in an attempt to protect the competitive position of indigenous companies. However, government policy can also reduce barriers to entry and increase the level of competition.

6.3.1.2 Supplier power

Bargaining power on the part of the supplier in an industry can be exercised through the increasing of prices or the rationing of supply. Clearly, the increasing of prices by powerful suppliers can reduce the profitability of the buyer. For example, the supplier power of Intel in microprocessors and Microsoft in operating systems has been a key influence on depressing the profitability of personal computer manufacturers. Characteristics of supplier power are the following.

- The purchase from the supplier group is important to the buyer.
- Suppliers provide a uniqueness or some level of differentiation in their product offering. Therefore, there are few alternative sources of supply for the buyer.

- There are high switching costs for the buyer. Switching costs are those costs that buyers must incur when changing suppliers.
- A buyer group is not an important customer to the supplier group. If the buyer group is an important customer, then the success of the supplier will be closely linked to the buyer group.
- The supplier poses a credible threat of integrating forward and acquiring the buyer. This potential threat of forward integration can depress the profitability of the buyer industry.

In effect, the key issues are the ease with which customers can switch supply sources and the relative bargaining power of both the buyer and supplier in the exchange.

6.3.1.3 Buyer power

Most of the issues dealt with under supplier power apply in reverse in the case of buyer power. For example, powerful buyers can suppress prices, demand higher quality and service levels, and play competitors off against one another, which in turn affects the profit potential of suppliers. The exercising of buyer power can be used as a source of competitive advantage for many companies. For example, many auto-makers such as General Motors and Ford have exercised buyer power over their suppliers in order to obtain higher service levels and lower prices. Characteristics of buyer power are the following.

- There are a limited number of buyers who purchase in large quantities.
- The products purchased by the buyer are standard with very limited differentiation. In other words, there are very low switching costs with the buyer having readily available alternative sources of supply.
- The suppliers that the buyer purchases from are highly fragmented.
- There are no information asymmetries in the buying process; i.e. the buyer has easy access to information on competing suppliers offerings and can use it as a source of bargaining power.
- The product being purchased by the buyer is non-critical and represents a small amount of the overall purchasing spend by the buyer.
- The buyer poses a credible threat of integrating backward and acquiring the supplier.

6.3.1.4 Substitute products

Substitute products are the alternatives on offer that can satisfy customer needs. The price that customers are prepared to pay for a product will be dependent upon the availability of substitute products. Furthermore, the availability of substitutes will have a significant influence on buyer power. For example, if there are readily available substitutes, not only between competing suppliers but also between competing products, then buyer power is likely to be high. The threat of substitutes is high when the following conditions hold.

- There are a number of cost-effective ways of satisfying the needs of the buyer.
- There are insignificant switching costs associated with switching supply sources.
- The buyer will switch to substitutes in response to price increases for the product i.e. demand is elastic in relation to price.

The extent to which substitutes can reduce prices and profits will be influenced by the ease with which customers can switch between alternative offerings. The strength of the substitute will depend upon whether it offers better value to the customer than existing products or services – whether it is cheaper, more durable, and convenient to use.

6.3.1.5 Rivalry

Rivalry is concerned with the intensity of competition within the industry. There is a high level of rivalry if competitors are continually reducing prices, introducing new products and advertising. There are a number of factors that indicate the intensity of competitive rivalry.

- There are numerous competitors in the industry that are of equal size and power. As the number of competitors increases, there is a greater likelihood of rivalry through actions such as price reductions and incentive offers to customers.
- If industry growth is slow or declining, then there are likely to be battles for market share. Competitors will attempt to maintain their market shares through price reductions or promotional activities. The intensity will be further increased if the product or service lacks differentiation and/or there are limited switching costs for customers.
- When there are high fixed costs relative to variable costs, companies will have to ensure that they are operating at or near to full capacity. This is evident in the case of the 'no-frills' airlines that will auction seats at very low prices in order to fill spare capacity on their flights.
- Exit barriers can affect the level of rivalry. Barriers to exit are the costs associated with leaving the industry. For example, government policy in the form of employment protection legislation can raise exit barriers. High exit barriers combined with excess capacity will have a detrimental effect upon industry profitability.
- In the case of an industry in which there are a diverse group of competitors in terms of origin, values and strategies, their actions are likely to be unpredictable. For example, the entry of the Japanese auto-makers into the US car industry led to an increased level of rivalry. Prior to their entry, competition was less intense with the US auto-makers enjoying a comfortable and stable environment. This was in part due to the similarities in strategies and mindsets that existed within these companies.

Analysis of the five forces can be used for the following purposes.

- *Positioning the company* – this involves positioning the company within the industry so that it can exploit its capabilities as a form of defence against its competitors. In effect, the company is aligning its strengths and weaknesses with the industry environment.
- *Influencing the balance* – this involves influencing the forces that drive competition. Rather than adopting a defensive stance, the company attempts to alter their stance in a way that will positively impact its competitive position. For example, if a company is the lowest-cost producer in the industry it may reduce its prices in order to eliminate weaker competitors, which in turn reduces the level of rivalry.
- *Exploiting industry change* – this involves exploiting industry changes as opportunities. For example, access to distribution channels is no longer a barrier to entry in many industries due to the development of the Internet. Companies such as Dell and Amazon have been able to sell directly to the end-customer via the Internet.

Understanding the forces that shape the competitive environment can serve as a basis for determining the capabilities that are likely to be critical for organisational success. The forces that shape the competitive environment are a key influence on the importance level of organisational activities. Some of the factors that influence the competitive forces are beyond the control of the company. For example, increasing de-regulation in many industries has meant that many companies have been unable to prevent the threat of foreign competitors. However, in some circumstances it is possible to influence some of these competitive forces in a way that will positively impact the competitive position of the organisation. For example, an organisation may redesign one of its products in order to differentiate it from competitive offerings. Such actions may allow the company to charge a premium price and reduce the level of rivalry with competitors. In this case, a strong design capability is a critical activity that can lead to competitive differentiation. The importance of activities can also change as the factors that influence the competitive forces change. Industries can evolve through competitors developing enhanced capabilities. This can lead to changes in the activities that traditionally have been a source of competitive advantage. For example, many of the large supermarkets have been investing considerable resource in the application of information technologies at point of sale in order to build rich profiles of consumer buying patterns. This information has become a strategic resource for these organisations and can be used for both marketing purposes and providing better forecasting information in their supply chains.

At a more general level, the five forces model can be used to understand the implications of outsourcing an activity or keeping it in-house. Choosing a particular sourcing option will have an impact upon one or a number of the five forces. For example, an organisation may decide to maintain control of the distribution network in an industry in order to limit the threat of new entrants. Alternatively, an outsourcing strategy may be employed in order to maintain the competitive position of the organisation in response to the actions of competitors. For example, an organisation may decide to outsource an activity in order to obtain the benefits of the scale economies or experience from a supplier already accessible to competitors or potential new entrants. The forces that influence competition will allow the organisation to identify the strengths and weaknesses in certain activities. These strengths and weaknesses can be determined from the causes of each force. For example, the threat of substitute products may indicate a weakness in innovation, which inhibits an organisation from differentiating its products through redesign or new product development. Choosing a particular sourcing option can be employed to improve the capability of the organisation in this area. The threat of new entrants and intense competitive rivalry can place significant pressures on organisations to reduce costs. Outsourcing can be used as a means of achieving such cost reductions. However, if outsourcing is being employed in order to reduce costs, it is important to be clear on whether the motive is to maintain competitive position or achieve a competitive advantage. Supplier power is another force that has to be carefully considered both in the evaluation and management of outsourcing. Mergers or acquisitions in the outsourcing supply market can lead to the emergence of a powerful supplier that can alter the balance of power in the relationship with the outsourcing organisation. A number of these issues will be further reflected upon in the Chapter 8 – *An Analysis of the Strategic Sourcing Options* – which deals with the implications of each strategic sourcing option.

6.4   Understanding the value concept

The value concept is widely used in a number of contexts including business strategy, marketing, finance and economics. Normann and Ramirez (1993) have argued that strategy is 'the art of creating value'. Strategy provides the intellectual frameworks, conceptual models and governing ideas that enable organisations to identify opportunities for creating value for customers and for delivering that value at a profit. Walters and Lancaster (1999) have offered definitions of three aspects of value.

(1) Value is determined by the utility combination of benefits delivered to the customer less the total costs of acquiring the delivered benefits. Therefore, value is a preferred combination of benefits compared with acquisition costs.

(2) Relative value is the perceived satisfaction obtained from alternative value offers available from other sources.

(3) A value proposition is a statement of how value is to be delivered to customers. It is important both internally and externally. Internally, it identifies the value

drivers it is attempting to offer a target customer group and the activities involved in producing the value together with the cost drivers involved in the value-producing activities. Externally, it is the means by which the firm positions itself in the minds of customers.

Value is also considered to be an important element of relationship marketing. The ability of a company to provide superior value to its customers is regarded as a successful competitive strategy. By adding more value to their products and services – through either quality improvements or support services – companies attempt to improve customer satisfaction in order to strengthen relationships and build customer loyalty (Ravald and Gronroos, 1996). Zeithaml (1988) identified a number of customer definitions of product value from the literature as follows:

(1) value is low price;
(2) value is whatever the customer wants in a product or service;
(3) value is the quality the customer gets for the price that is paid; and
(4) value is what the customer gets for what he/she gives.

Utility theory is the theoretical basis of the concept of value (Lancaster, 1971). Utility theory emphasises that customers do not purchase products or services for their own sake. Rather, customers purchase bundles of attributes that together represent a certain level of service quality that is offered by a company at a certain price level. In other words, customers do not buy products; they buy a set of expected benefits. These benefits may be intangible. Rather than relate to specific product features, they are concerned with issues such as image and/or reputation. Monroe (1991) defines customer-perceived value as the ratio of perceived benefits relative to the perceived sacrifice. The perceived sacrifice encompasses all the costs the customer has to make when making a purchase. These costs include purchase price, acquisition costs, transportation, installation, order handling, repairs and maintenance, risk of failure or poor performance. The perceived benefits are some combination of physical attributes, service attributes and technical support available in relation to the particular use of the product, as well as the purchase price and other indicators of perceived quality. A key task for organisations is to increase the value provided to the customer. Using Monroe's (1991) definition of customer-perceived value, Ravald and Gronroos (1996) argue that value can be increased in one of the following ways:

- *Increasing the benefits* – through adding something to the product that the customer perceives as important, beneficial and of unique value. For example, high product quality plus supporting services can increase the benefits for the customer which in turn positively impact customer-perceived quality.
- *Reducing the sacrifice* – this is concerned with examining how a company can add value to the offering by reducing the customer-perceived sacrifice. Reducing the sacrifice the customer has to undertake in order to purchase a product may involve reducing the price of the product or increasing the convenience of the purchase.

In effect, both the benefits and sacrifices are mutually reinforcing. Increasing the benefits should lead to a reduction in the customer-perceived sacrifice through a reduction of the costs involved in the purchase. Clearly, the creation of value has significant implications for competitive advantage. Webster (1994) has argued that strategy must be based on an analysis of the company, the competition, and the customer, identifying those opportunities for the firm to deliver superior value to customers based on its distinctive competencies. Unless the product offered by the organisation is distinct in some way from competitors, it is likely that customers will view the product as a commodity and purchase primarily on the basis of price. Therefore, it is crucial that the organisation attempts to add 'additional values' to the product in order to differentiate it from competitors. In order to do this, the following questions should be answered.

- What are the features in the product that create value for the customer?
- What is the customer actually paying for?
- Why is the customer willing to pay more or less for one product or service than a competing offering?
- Which value features are most important to customers and thus make the most significant contribution to price?

A useful approach for understanding value is the 'perceived use value' approach proposed by Bowman and Faulkner (1997). They define the value of a product or service as the perceptions that a customer has of the usefulness of the product. In determining value, customers are assessing the extent to which their needs are being satisfied by purchasing a product. Total value is the amount that the customer is willing to pay for the product. Total value is made up of two types of perceived use value.

- *Hygiene value* – refers to those elements of the product that all competitors offer. These are the standard order-qualifying elements that are a pre-requisite for participation in the market.
- *Motivator value* – refers to the elements of the product or service that enable the company to win customers. These dimensions of perceived use value are not offered by all companies and are generally unique to a particular company. Therefore, these motivator elements can be a source of differentiation.

Companies should focus on increasing the level of motivator value to customers, which in turn will enhance its position in relation to competitors. In order to maintain a sustainable competitive position, companies should attempt to identify the perceived use value that is likely to act as a source of competitive advantage in the future. Bowman and Faulkner (1997) argue that understanding the customer's definition of value is central to competitive strategy. Competition is driven by the buying behaviour of customers. The competitive position of an organisation can be assessed by understanding the buyer behaviour of customers. Total value can be estimated by determining the amount customers are prepared to pay for a product. In a market where there is a high level of competitive rivalry and a wide range of choice, customers can capture most of the value on offer. The wide range of choice indicates a low level of differentiation, which in turn shows that the companies in the market are providing a very small amount of motivator value. However, in a market where there is a low level of competitive rivalry and customer choice the company can capture a significant part of the total value. Futhermore, in this case motivator value is a significant part of the total value. Motivator and hygiene value have implications for outsourcing evaluation and management. Activities that are closely linked to motivator factors are likely to be considered critical, whilst activities that are closely linked to hygiene factors are non-critical. Motivator value must be continually monitored in response to changes in the business environment. For example, over time the factors that contribute to motivator value may become hygiene factors due to a number of influences such as changing customer preferences or developments in the capabilities of competitors. Therefore, the importance of certain critical activities linked to motivator factors may diminish, as these same motivator factors become hygiene factors. Bowman and Faulkner (1997) have suggested that the following steps should be carried out in the application of this approach.

6.4.1   Clarify the target market

This involves defining a target market. This target market can be a specific customer or type of customer with certain requirements. In fact, an understanding of customer requirements is a crucial starting point in determining the competitive position of the organisation. BMW has designed specific models for different income and age groups. It sells models for target markets with varied combinations of ages and income. For example, the short wheel-base 3 Series is targeted at young 'upwardly mobile' individuals whilst the 7 Series is targeted at the more mature executive-type individual. A group of customers are likely to have similar requirements. These requirements can be identified in order to define the perceptions of use value. A useful approach is to identify one individual that is a fair representation of the needs of all within the target market. It is possible to derive target markets on the basis of the following:

- *Customer and personal characteristics* – these can include age, income, sex, location, lifestyle and social class. These types of characteristics can be determined from traditional market research used to determine market segments.
- *Purchase and usage behaviour* – is concerned with the nature of the interaction between the customer and the product. For example, a company may have to determine how, when, where and the volume in which the product is purchased. In this way target markets or segments can be categorised on the basis of where customers purchase the product, and frequency and level of purchase. For

example, the needs of customers who purchase hi-fi equipment in high street multiples are likely to different from those who purchase in specialist hi-fi stores.

- *Customer needs and preferences* – involves an in-depth understanding of customer behaviour and objectives. This is concerned with understanding what motivates customers, the expected benefits they are seeking, and the nature of their preferences. For example, ready-cooked meals are an interesting illustration of the different needs of customers expect from a product. Some customers may buy these products as a quick convenient product whereas others perceive them as a substitute for preparing the same meal at home.

6.4.2   Identify the dimensions of perceived use value

This step can be accomplished through determining customer needs and how customers evaluate the products. For example, consider the case of a high technology company in the electronics industry designing and manufacturing complex electronic components for multinational companies. The basic needs of the customers of this company are the design, manufacture and delivery of high-quality, technologically sophisticated components at competitive prices. These basic needs can be translated into a set of dimensions of use value including design capability, time-to-market, quality levels, delivery service and manufacturing capability. Although in this example there are many factors that are valued by customers (comparable to hygiene factors), these dimensions of perceived use value are deemed to be the most important. The company's customers are seeking a set of perceived use values that the company must fulfil in order to meet their basic needs.

6.4.3   Weight the importance of the dimensions of perceived use value

This step is concerned with ranking the importance of the selected dimensions of perceived use value. The dimensions should then be weighted using percentages. This is shown in Figure 6.1 that illustrates the example of the high technology company.

6.4.4   Rate competing products on each value dimension

This stage is concerned with rating the products of competitors along the dimensions of perceived use value. This is also shown in Figure 6.1.

6.4.5   Determine the perceived price

This stage is concerned with determining the price factors. Price factors can include the price of the product along with the costs associated with owning and

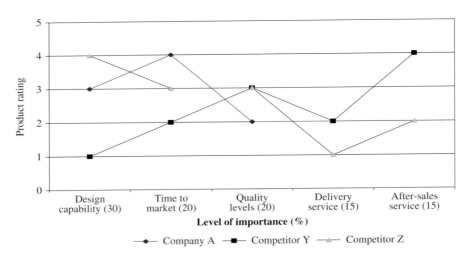

Figure 6.1 Samples of perceived use values for high technology company

maintaining the product. For example, when someone purchases a car, they are not only concerned with initial price, but are also concerned with fuel consumption, insurance costs and maintenance, and the costs associated with switching from one company to another can also be included. Therefore, switching costs can increase the perceived prices of competitors' products.

6.4.6   Plot the products on the customer matrix

Once the perceived use value and perceived price have been determined, the product can be plotted on the matrix. Figure 6.2 shows a sample matrix representing the position of a company and a number of its competitors. The customer matrix is derived from the perceptions that customers have of products or services on offer and the price being charged. The vertical axis refers to the perceived value obtained by the buyer in purchasing and using the product or service, and the horizontal axis represents the perceived price. Perceived use value and perceived price are the two components of value for money. Perceived price includes the elements of price with which the customer is concerned. Perceived use value refers to the benefits obtained by the customer from purchasing the product. The product positions are shown as points that represent the perceptions that an average customer has of each of the products perceived to be on offer.

In terms of the implications of the position of products on the matrix, potential options include reducing the perceived price, adding to the level of perceived use value, or doing both. Pursuing a price reduction strategy will involve reducing costs across the business, whilst adding to the level of perceived use value will involve focusing resources on activities that differentiate the products and services

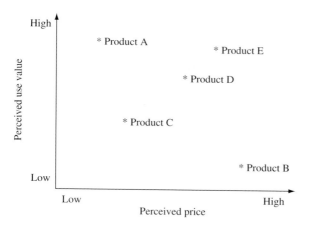

Figure 6.2 Sample products on customer matrix

of the organisation in the eyes of its customers. The potential options of reducing price or adding to the level of perceived use value are influenced by Porter's generic strategies of cost leadership or differentiation. The positions in relation to perceived price and perceived use value can be used to determine both the relative importance of organisational activities and provide guidance on the most appropriate sourcing option. The importance of an activity will be significantly influenced by its impact upon the ability to reduce perceived price or the dimensions of perceived use value. In relation to perceived price, any activities that allow the organisation to realise substantial cost reductions should be considered. However, the basis of the any cost reductions realised should be understood. Although an activity may be a considerable proportion of the cost of the product or service, it does not automatically follow that it is critical. For example, in a manufacturing context the most significant cost in the end product is that of manufacture particularly if it is highly labour-intensive. In many instances, such activities can be readily sourced by the organisation and its competitors from a number of more competent suppliers with lower costs. Therefore, the sourcing option in this case only maintains the competitive position of the organisation. However, if the organisation can realise significant cost reductions in an activity by possessing a distinctive capability in the activity, which is based on, for example, either scale economies or experience effects, then this is likely to be a more critical activity. In relation to perceived use value, activities that make a significant contribution to the most important dimensions of perceived use value are likely to be critical to organisation success. These are likely to involve creative activities such as product design and brand management, which allow the organisation to differentiate itself from its competitors.

By using this approach to exploring customer needs and perceptions, it is possible for an organisation to determine what is valued in their products and services and how outsourcing can be used as a vehicle to either reduce perceived price or enhance perceived use value. For example, outsourcing an activity to a more competent supplier may lead to a reduction in costs, which can make a positive contribution to reducing the perceived price. Reducing the perceived price whilst providing the same of level of perceived use value may lead to an increase in market share. However, the potential reaction of competitors to this action must be considered. For example, reducing the price may lead to a price war as competitors attempt to protect their market share. It is also possible that by reducing price there is a negative impact on customer perceptions of perceived use value. For example, in the case of luxury-type products reducing price may cheapen the product in the eyes of the customer. Therefore, it is important when making a decision on price that there is a clear understanding of why the customer purchases the product. In order to compete on price, it is crucial that the company has a lower cost base than its competitors in order to be able to continually reduce price and sustain prices for longer than competitors. This is likely to be a continuous process of pursuing cost reductions across the entire business including scale economies, rapid new product introductions and access to a competitive supply base.

Choosing the most appropriate sourcing option can also add to the level of perceived use value. For example, investing in an activity performed internally that contributes to an important dimension of perceived use value may allow the company to differentiate itself from its competitors. In order to impact this dimension, a thorough understanding of the customer and how the customer evaluates the product is required. Essentially, the company is attempting to positively influence the dimensions of perceived use value in order to enhance its competitive position. Referring back to the high technology company in the electronics industry and the dimensions of perceived use value on Figure 6.1, it can be seen that design capability is perceived by customers to be the most important purchasing dimension. Currently, Company A has a slightly weaker position in design capability in the eyes of the customer in relation to Competitor Z. A potential strategic option for Company A is to further develop its capability in this area in order to match or surpass the position of Competitor Z in this area. It is also important to consider the overall strengths that both the company and its competitors possess. For example, considering the position of Competitor Y, it can be seen that its strengths do not reside in the areas that are most valued by customers, whereas the strengths of Company A are more closely aligned with customer needs. Moreover, customers' perceptions of value can change over time. For example, a competitor that improves its performance over time in a way that positively impacts the dimensions of perceived use value can harm the position of other companies in the market.

Both dimensions of perceived price and perceived use value should be combined with an analysis of the capability of the sourcing organisation in certain activities.

Combining the analysis of customer needs and perceptions with capability analysis will influence the choice of the most appropriate sourcing option. For example, ideally an activity that makes a significant contribution to a strategy of increasing perceived use value and in which the organisation possesses a particular performance advantage over competitors should be kept in-house and further developed. In addition, potential strategy options that can arise from this analysis should also be considered in the context of the five forces that shape industry structure. For example, in an industrial buying situation any attempts to enhance the level of perceived use value will create little benefit for the selling organisation if powerful buyers capture most of the value that is created. Furthermore, the likely implications of a strategy option on the forces that shape industry structure and in particular the capabilities required to counteract competitive reactions must be considered. For example, an organisation should possess strengths in cost reduction if it attempts to reduce the perceived price because of the potential for increased competitive rivalry and the migration of competition towards price. The sustainability of a perceived price reduction or perceived use value strategy has to be considered – either through outsourcing or building a capability internally. For example, if the company decides to outsource an activity to a supplier in order to improve a dimension of perceived use value then it has to consider the ease with which competitors can do likewise. If the same supplier is accessible to competitors then it is unlikely that such a position can be sustained over time. In order to sustain such a strategy, the company may consider developing a unique relationship with the supplier that is difficult for competitors to replicate. In the case of investing internally to develop a capability, the company has to consider its current capability in relation to competitors and the ease with which competitors can replicate such a strategy.

6.5   The critical success factors methodology

Critical success factors (CSFs) are those things that the organisation must perform well in order to ensure success. CSFs represent those areas of the organisation that management needs to focus on to create high levels of performance. Rockart (1979) was the first to define the concept of CSFs as the limited numbers of areas in which results, if they are satisfactory, will ensure competitive performance for the organisation. The CSF approach is a top-down methodology to assist business strategy development. As well as identifying the CSFs, it also highlights the key information requirements for senior management (Byers and Blume, 1994). Moreover, if the CSFs are identified and controllable, management can take appropriate action to improve the potential for organisational success. CSFs include issues vital to an organisation's current operating activities and to its future success. Therefore, the

CSF methodology attempts to identify the few key areas that are most likely to lead to organisational success. Examples of CSFs include the development of a global sales network, superior new product development and the development of a best-in-class supply base. In most organisations, between four and eight CSFs can be identified as crucial to the performance of the organisation. The sources of CSFs will be influenced by a number of factors including industry characteristics, stage of the industry value chain, and the needs of customers groups. Many industries have a unique set of CSFs that are determined by industry characteristics. For example, CSFs for retailers can include location, inventory turnover, promotional activities and price. An important issue when considering CSFs at the industry level is that the CSFs change as the industry changes. The use of CSFs at an industry level may be dangerous if the entire industry is under-performing and is losing its competitiveness as other industries better serve customer needs through substitute products and services. Differing stages in an industry value chain may also influence the choice of CSFs. For example, the success of a retail business is heavily influenced by store location. In contrast, a wholesaler selling to this retailer would not normally consider location to be a CSF. Factors that are particularly important to certain customer groups can also influence the choice of CSFs. For example, in a business-to-business context many suppliers to multinational corporations consider factors such as flexible delivery services and reduced lead times to be critical to successfully serving the needs of this group of customers.

Illustration 6.2

Developing critical success factors

Paula van Veen-Dirks and Martin Wijn provide a number of case studies of the development of critical success factors (CSFs) for a number of companies in their work on integrating the CSF method with the balanced scorecard. In particular, they provide an interesting illustration of a Dutch retail group and how it determined CSFs for its business. The company had a network of around 120 retail outlets in the Netherlands, Belgium, Germany and the UK. Its product range included accessories for cars and bikes and mobile communication. Although the company was the leader in its home market of the Netherlands, it was experiencing considerable competition across its business areas. Whereas the company was not competing in the upper end of the market, it was not acting as a discounter. The company defined its mission statement as: 'The best offer for people on the move'. The company was pursuing a strategy which involved focusing on a number of key areas including the following.

- It was attempting to strengthen its dominant market position in a number of countries by integrating its existing and newly acquired stores and disposing of more than 100 stores.
- The company was striving to reduce costs across all its business activities.
- Its 'retail formula' was based on both 'competence' and 'dominance'. Competence refers to the capability of the company to provide the range of products and services on offer. By achieving dominance the company believed that the customer would buy their products and not consider the other options available.

In the buying decision, customers were motivated primarily by the price/quality performance dimension. However, customers were also looking for a 'fun shopping experience' when buying the product in the retail outlet. Due to rapidly changing buyer behaviour the company decided to develop a number of market-oriented critical success factors which included the following:

- knowledge of the customer;
- product range;
- price;
- attractiveness of the retail outlets; and
- the presence of skilled/helpful employees.

The company used these CSFs at corporate level for its range of products and services. In order to achieve its strategic objectives high levels of performance were required in each of these CSFs. It was also possible to use these CSFs as a basis for identifying critical business activities. Critical business activities were selected on the basis of having the highest impact upon the level of performance of the CSFs. The company chose the following as critical business activities:

- informing and serving the customer;
- selecting new products;
- determining the selling price;
- developing the retail formula; and
- recruiting and selecting personnel.

The determination of recruiting and selecting personnel as a critical activity reflected the choice of the presence of skilled/helpful employees as a CSF. For example, the analysis undertaken emphasised the importance of human resource management to the success of the company. Conditions in the labour market such as increasing labour costs and high labour turnover had made it difficult to recruit high-quality personnel. In addition, the trend towards extended opening hours across all its retail outlets led to lower sales per hour and increases in labour costs.

Source: van Veen-Dirks, P. and Wijn, M. (2002). Strategic control: meshing critical success factors with the balanced scorecard. *Long Range Planning*, **35**, 407–27.

Van Veen-Dirks and Wijn (2002) have found that the CSFs methodology can help practitioners deal effectively with the tension between strategy implementation and formulation. CSFs are also regarded as a means of selecting non-financial indicators. As well as being used in a business strategy context, the method can be easily adapted to a number of different applications including information systems development and business improvement. The CSFs methodology is valuable in outsourcing evaluation and management in that it can establish a direct link between the potential sourcing options and the market. For example, market-oriented CSFs are a very effective tool for providing information about customer needs and potential sources of competitive differentiation. The CSF method provides a language that generates acceptance at the senior managerial level. It also can serve as a means of identifying important areas for strategic attention. Using performance measures to monitor the status of CSFs, changes in the market can be detected directly allowing appropriate management action to be taken. Examples of performance measures associated with CSFs are shown in Table 6.1. When changes are identified, the monitoring system can flag them as part of the business strategy process and highlight the need for changes in strategy.

There are two key elements of the CSF methodology.

6.5.1   Identification of CSFs

The CSFs can be developed from both the mission statement of the organisation and from an evaluation of the organisation's competitive position. Hardaker and Ward (1987) argue that a sound basis for the identification of CSFs is to consider the mission statement of the organisation. The mission statement of an organisation is a general statement of the purposes of an organisation and should express the over-riding raison d'etre of the business. Mission statements are built around a number of elements including a statement of the overall mission of the company; a summary of the key values that managers are committed to and that influence the management of the organisation; and the articulation of the key goals that management believes must be adhered to achieve the mission (Hill and Jones, 2001). Using the mission statement as a basis, CSFs can be defined on the basis of what the organisation has to accomplish in order to achieve its mission. The competitive environment also has a significant impact upon the determination of CSFs. For example, the level of competitive rivalry along with the threat of new entrants is likely to have an impact upon the CSFs chosen. In industries where there is a high level of competitive rivalry, success can be heavily dependent upon pursuing aggressive cost reduction strategies. The use of Porter's five forces framework can assist in this analysis. The analysis of each force and the inter-relationships between them can provide important information to assist in the identification and justification of CSFs.

Table 6.1 *Performance measures associated with critical success factors*

Critical success factors	Performance measures
Profitability	Profit measures across products and territories. Comparisons between products and territories as well as competitors.
Marketing capability	Relative market share, sales growth, marketing expenditure/sales (%). Sales comparisons between products and territories as well as competitors.
Human resource capability	Revenue per employee, staff turnover rates, rate of absenteeism, average length of absence per employee, number of long-term vacancies, etc.
New product development capability	Research and development/sales (%), time-to-market, rates of new product introduction (NPI), new product sales/overall sales (%). Comparisons of these measures between competitors.
Financial strength	Financial ratios including liquidity, gearing, profit and efficiency.

Consideration should also be given to customer preferences in relation to the products and services of both the organisation and its competitors. This involves identifying both who the customers are and understanding why they purchase a product or service from the organisation or one of its competitors. For example, in the clothing industry retailers are often willing to pay a premium price for specialist fashion items if suppliers can produce exclusive designs and provide reliable delivery service and the rapid replenishment of fast-moving items. This contrasts with the food industry where retailers are primarily concerned with obtaining the lowest price for commodity-type items such as meat and dairy produce. Capabilities in certain activities can also influence the identification of CSFs. For example, an organisation may decide to further invest in an activity in which it has particular strength because the activity is likely to become a key factor for success in the future. Clearly, CSFs can be derived from a combination of factors in both the external and internal environment. For example, a key determinant of success in the semiconductor industry is the rapid introduction of new components with higher levels of functionality and at a lower cost. Many of the customers of these companies are personal computer manufacturers such as Dell and Gateway who, as well as demanding the rapid introduction of more advanced components, are also requiring shorter lead-times for components in order to reduce the risks of component obsolescence. The industry is also characterised by intense competition and component life cycles that are as short as a few months. Companies in the semiconductor industry must invest vast resources in the design and manufacture of components technologies and develop logistics skills that facilitate shorter delivery times and higher levels of service for their customers. Therefore, rapid new product development capabilities and logistics skills are likely to be critical factors for success for companies in this industry.

6.5.2 Determine performance measures to monitor CSFs

Performance measures should be determined to allow the organisation to monitor progress on whether the critical success factors are being met. For example, a retailer may identify a CSF as the achievement of supply chain excellence. In order to monitor progress on this CSF, performance measures may include order management costs; total logistics costs; average inventory days of supply; and days of sales outstanding. CSFs can also assist in determining what information is required in order to monitor the overall performance of the organisation. After the measures have been determined, the most appropriate ones must be selected. As part of this analysis the following questions should be answered.

- Can each measure be externally benchmarked?
- Can each measure be linked to the business strategy of the organisation?
- Will there be an understanding of each measure across the organisation?
- Can each measure be easily calculated, recorded and reported?

Appropriate levels of information must be collected in this process. This may involve a review of budgets, business plans, and planning documents to identify current CSFs. Senior management along with participation from lower levels may discuss the company's strategic goals and formulate new critical success factors. CSFs can emerge from structured dialogues amongst the key personnel within the organisation. A series of dialogues with managers should result in an explicit statement disclosing each individual's personal CSFs. It is important that CSFs are elicited from managers who represent a cross section of the organisation's major functional areas. This will provide a collection of consistently referenced CSFs that can be extracted and refined into a set of organisational CSFs.

Once the CSFs have been determined, they can be used as a basis for determining the importance level of activities for outsourcing evaluation. The CSFs – as determined by an analysis of the competitive environment and customer needs – should translate into a number of critical activities that create value for the organisation. The first step in this analysis involves determining how specific activities contribute to the achievement of the CSFs. For example, a CSF such as 'the operation of a cost-effective and safe manufacturing operation' needs to be clearly defined in terms of what it exactly means. In effect, the organisation is identifying what has to be done in order to meet the CSFs that are central to its current and future success. This analysis can be carried out by using either the business process or the value chain approach. In effect, the organisation is taking the analysis carried out in Chapter 5 and linking the activities to the CSFs identified. Many of the activities identified will have some impact upon the CSFs. However, it is necessary to distinguish between the importance level of each activity. One potential approach is to map the activities and their impact upon each CSF. The process should begin by identifying the activities that impact upon each CSF. It is likely that one activity will impact upon a number of CSFs. A reliable indicator of the importance is the number of CSFs the activity impacts. This methodology is illustrated in the case of a retail bank in the following case illustration.

Illustration 6.3

Critical success factors and critical activities in the financial services sector
This case illustration provides an overview of the CSFs developed by a retail bank in the UK. It also outlines how critical activities were identified from these CSFs. The retail bank established the following CSFs.

- *Reduce the costs of routine transactions* – refers to the capability of the bank to manage cash and paper transactions (debits and credits on behalf of their customers). The capability to process credit applications was also an important element of this capability. Some of the activities related to this CSF included counter management in branches, ATMs, Internet banking, customer credit application interviews, and security perfection.
- *Enhance sales and service capability* – a critical element of the bank's strategy had been to decentralise many of the administrative activities in branches and focus more resource on direct customer sales services such as credit assessment and delivery, financial advice, and share dealing. Some of the activities central to this CSF included customer relationship management, research of the marketplace and credit assessment.
- *The development and retention of personnel* – was central to achieving its business strategy. Activities central to achieving this CSF included recruitment, training and development, personal development programmes, and staff attitude surveys.
- *The development of innovative products and services* – the need for new and innovative products was increasing in the financial services sector. Also, the high level of regulation in the financial services sector means that new products have to be tested rigorously before release to the market. In the development of new products, the bank could either develop new products internally or form strategic alliances in order to enhance its product and service offering. Important activities for this CSF included research of the marketplace and customer relationship management.

- *The management of customer information as a strategic asset* – the trend towards a greater focus on building customer relationships has led to the need for a greater understanding of customer behaviour and needs. The bank requires a greater knowledge of what products their customers currently use, products that their current customers are buying from competitors and the products that their customers are unaware that the bank can provide. Some of the activities central to this CSF included analysis of customer satisfaction surveys, advertising and creating awareness, and understanding customer requirements.
- *Governance and regulation* – the Financial Services Authority (FSA) regulates the financial services sector in order to prevent malpractice in a number of areas including money laundering and offering of poor advice. Compliance with the FSA involves acting fairly in dealings with customers and assisting customers in understanding how their financial products and services work. Compliance with these regulations was perceived as enhancing the brand image of the bank. Providing advice, confidentiality, and excellent customer service are central to building this brand. Important activities for this CSF included compliance with FSA guidelines on 'Best Advice' and 'Money Laundering', keeping up to date on legislation and adherence to the Banking Codes.

The overview of the CSFs identified has shown that some of the business activities were common to a number of CSFs. Using Ward's (1990) methodology, further analysis revealed the level of criticality of each activity based upon its relationship with the CSFs as shown in Figure 6.3. It is possible to use this method for assessing the criticality of each activity. Two variables can be used including *the importance of the activity* and *the quality of the activity*. The number of times an activity is considered critical to the achievement of a CSF is summed in column (A). Activity quality in column (B) refers to the current performance of the organisation in the activity concerned ranging from 'A' – no need for improvement to 'E' – still has to be developed. Ward's (1990) methodology regards critical activities as those that are central to the future success of the company as indicated by the count scores on the importance for the CSFs.

However, this approach should be applied with care and in particular should be combined with the knowledge and experience of key decision-makers within the organisation. Determining the importance level of organisational activities is a complex task, and care must be taken to ensure the long-term strategic considerations and true benefits are assessed. This process of identifying the importance level of activities should be carried out by senior management along with inputs from teams from lower levels in the organisation. It is important that the

Activities \ Critical success factors	Reduce the costs of routine transactions	Enhance sales and service capability	The development and retention of personnel	The development of innovative products and services	The management of customer information as a strategic asset	Governance and regulation	(A) Count	(B) Activity quality
Money transmission	X	X	X	X	X	X	6	D
Credit delivery e.g. mortgages and securities	X	X	X	X	X	X	6	E
Communication systems		X	X	X	X	X	5	B
Research and understand the marketplace		X		X	X	X	4	B
Selling skills		X	X	X	X	X	5	C
Customer satisfaction surveys		X		X			2	C
Updating customer profiles		X			X		2	C
Understanding customer requirements		X	X	X			3	C
Credit assessment skills		X	X	X	X	X	5	A
Advertising		X		X			2	B
Customer relationship management		X	X	X	X		4	C
Recruitment		X	X				2	B
Training and development		X	X	X	X	X	5	B
Remuneration/retention		X		X			2	A
Undertaking staff attitude surveys			X				1	A
Compliance with FSA guidelines						X	1	B
Adherence to banking codes						X	1	A

Figure 6.3 Sample critical success factors and activities matrix for a retail bank

participants in the process work closely in group sessions to ensure that strong debate takes place. Group consensus building techniques such as decision analysis or decision conferencing techniques could be employed.

The use of the CSF method as a tool in business strategy has received criticism from a number of authors. For example, Ghemawat (1991) has argued that it is difficult to determine critical success factors because there are so many factors with complex interdependencies that determine success. The complex and dynamic nature of many business environments also makes it extremely difficult to generalise and identify a limited set of factors that are likely to ensure business success. Some of these criticisms could also be levelled at attempting to understand the concept of value and the competitive environment. However, in outsourcing evaluation and management it is important to place these techniques in their proper context. The objectives of the CSF method can be easily understood. The CSF method focuses attention on a set of key issues, and then proceeds to refine the issues in a way that allows scrutiny for validity and completeness on an ongoing basis. CSFs provide a means of linking outsourcing with the key factors that are likely to determine success in the business environment. For example, an organisation may decide to outsource an activity previously carried out in-house because it no longer makes a significant impact upon the attainment of a CSF. In relation to the value concept, building a capability in a particular process with a supplier may have a positive impact upon an important dimension of the product or service that the customer values. Although it is difficult to predict future changes in the business environment, the dimensions of the perceived use value can be refined to reflect any changes that occur in the business environment. Even though there is uncertainty and complexity in many business environments, it is important to have some mechanism that attempts to make sense of the business environment even though these techniques have their limitations. Customer preferences and the competitive environment are key influences on outsourcing evaluation. In particular, knowing the importance level of activities enables an organisation to determine whether outsourcing can be used to maintain the competitive position of the organisation or act as a source of competitive advantage.

REFERENCES

Byers, C. R. and Blume, D. (1994). Tying critical success factors to systems development. *Information and Management*, **26**, No. 1, 51–61.

Bowman, C. and Faulkner, D. (1997). *Competitive and Corporate Strategy*. London: Irwin.

Ghemawat, P. (1991). *Commitment: The Dynamic of Strategy*. New York: Free Press.

Hardaker, M. and Ward, B. K. (1987). Getting things done. *Harvard Business Review*, **65**, No. 6, 112–20.

Hill, C. and Jones, G. R. (2001). *Strategic Management: An Integrated Approach*. Boston: Houghton Mifflin Company.

Kroll, M., Wright, P. and Heiens, R. A. (1999). The contribution of product quality to competitive advantage: impacts upon systematic variance and unexplained variance in returns. *Strategic Management Journal*, **21**, 375–84.

Lancaster, K. (1971). *Consumer Demand: A New Approach*. New York: Columbia University Press.

Miller, A. and Dess, G. G. (1993). Assessing Porter's (1980) model in terms of its generalisability, accuracy and simplicity. *Journal of Management Studies*, **30**, No. 4, 553–85.

Monroe, K. B. (1991). *Pricing – Making Profitable Decisions*. New York: McGraw-Hill.

Normann, R. and Ramirez, R. (1993). From value chain to value constellation: designing interactive strategy. *Harvard Business Review*, **71**, No. 4, 65–77.

Porter, M. E. (1980). *Competitive Strategy: Techniques for Analysing Industries and Competitors*. New York: Free Press.

Porter, M. E. (1985). *Competitive Advantage: Creating and Sustaining Superior Performance.* New York: Free Press.

Ravald, A. and Gronroos, C. (1996). The value concept and relationship marketing. *European Journal of Marketing*, **30**, No. 2, 19–30.

Rockart, J. F. (1979). Chief executives define their own data needs. *Harvard Business Review*, **57**, No. 2, 81–92.

Van Veen-Dirks, P. and Wijn, M. (2002). Strategic control: meshing critical success factors with the balanced scorecard. *Long Range Planning*, **35**, 407–27.

Walters, D. and Lancaster, G. (1999). Value and information – concepts and issues for management. *Management Decision*, **37**, No. 8, 643–56.

Ward, B. (1990). Planning for profit. In: *Managing Information Systems for Profit*, Lincoln, T. (ed.). Chichester: Wiley.

Webster, F.E. (1994). *Market Driven Management*. New York: Wiley.

Zeithaml, V.A. (1988). Consumer perceptions of price, quality and value: a means-end model and synthesis of evidence. *Journal of Marketing*, **52**, July, 2–22.

7 Capability analysis

7.1 Introduction

A major part of outsourcing evaluation is determining whether an organisation can achieve superior performance levels internally in critical activities on an on-going basis. Clearly, if the organisation can perform the activity uniquely well, then this activity should continue to be carried out internally. However, many organisations assume that because they have always performed the activity internally, then it should remain that way. In many cases, closer analysis may reveal a significant disparity between their internal capabilities and those of the world best suppliers and competitors. Organisations considering outsourcing must rigorously evaluate their capabilities in relation to both their suppliers and competitors. Analysis from an activity perspective is important given that organisations are increasingly positioning themselves in specific parts of the value chain in order to gain competitive advantage. In effect, each organisation is in competition with all potential suppliers of each activity in its value chain. In the case of a critical activity, each point in the value chain relevant to providing the activity should be assessed. Again, this analysis is concerned with identifying the disparity between the sourcing organisation and potential external sources. This analysis can identify sources of competitive advantage that can be exploited more fully by further developing certain activities. It also assists in revealing weaknesses that need to be addressed – either through internal improvement or outsourcing – to become more competitive. It can allow an organisation to focus on whether it will be detrimental to their competitive position to outsource activities such as research and development, design, engineering, manufacturing, or assembly, both in the short- and long-term. The benefits of carrying out this analysis are as follows.

- Organisations can focus resources on activities where they can achieve pre-eminence and provide unique customer perceived value. If an organisation has leadership in an activity considered to be a key source of competitive advantage then this activity should be held and further developed within the organisation in order to maintain and build upon this position. For example, in the 1990s Eastman Kodak had world leadership in two of its critical activities – chemical imaging and electronic imaging. Thus, these activities were held within the company in order to maintain and build upon this leadership (Hamel and Prahalad, 1994).

- Activities for which the organisation has neither a critical strategic need nor distinctive capabilities are potential candidates for outsourcing. It is more appropriate for an organisation to use external suppliers that are more competent and have a lower cost base. For example, Unilever developed its own advertising function internally for many years before divesting this agency. Pursuing this strategy allowed Unilever to have access to whichever firm it considered most appropriate for the task in hand (Alexander and Young, 1996).

Determining the capability of an organisation in relation to competitors or suppliers in the context of outsourcing involves an analysis of the following:

7.1.1   The type of advantage

The advantage can be based on attributes such as lower costs, superior quality, service levels, etc. Superior performance in an activity can also include a combination of these attributes. There are a number of elements of this analysis.

- *Cost analysis* – part of this analysis involves comparing the costs of sourcing the activity internally and from an external supplier. An assessment of the relative cost position of the sourcing organisation in relation to both suppliers and competitors in the activities under scrutiny should also be undertaken. An assessment of costs can form a significant part of capability analysis. The major drivers of cost associated with each activity should be identified. For example, for capital-intensive activities the major cost drivers are likely to be the cost of equipment and production volume. Alternatively, in highly labour-intensive manufacturing processes, the major driver of costs is labour rates.
- *Benchmarking* – can assist in determining performance levels in the activities under scrutiny. Organisations considering outsourcing must rigorously evaluate their capabilities in relation to suppliers and competitors. This analysis involves a structured benchmarking approach to assessing the organisation's capabilities relative to potential suppliers and competitors. Benchmarking also involves consideration of the cost position relative to competitors and suppliers.

7.1.2   The source of the advantage

This part of the analysis is concerned with understanding how superior performance in the activity is achieved. Determining the type of advantage will have revealed valuable insights into the reasons for any disparity in performance identified. For example, the analysis may have revealed that external suppliers can provide the activity at a much lower cost than internally within the sourcing organisation. However, it is important to understand how this relative cost position is achieved. The superior cost performance of the external suppliers may have been as a result of scale economies and greater experience in the activity under scrutiny. These types of insights will inform the analysis of the sustainability of a superior performance position. This type of analysis can also assist in understanding how performance can be improved internally without having to outsource the activity. Many organisations mistakenly rush into outsourcing without attempting to understand the underlying causes of poor performance.

The level of analysis undertaken in relation to capability analysis will depend upon the level of importance of the activity. Activities identified as being highly important will require extensive analysis of performance levels in relation to both competitors and suppliers. The basic nature of strategic analysis may move from an industry analysis to a horizontal analysis of capabilities across all potential providers of the activity regardless of which industry the provider might be in. For example, these activities may include logistics, warehousing, process design, product design, quality control and plant engineering as well as the actual production that may be a function of the workforce skills. As well as the importance of the activity under scrutiny, the manner in which capability analysis is undertaken will also be influenced by the priorities of the sourcing organisation. These priorities may have been determined on the basis of the following.

- The organisation may have identified a number of critical activities where it feels its performance is lacking in comparison to external sources. Therefore, urgent action is required in order to determine the significance of this disparity in performance.
- The organisation may have identified a number of less important activities that it considers as potential candidates for outsourcing.
- Organisational difficulties in certain areas may require rapid performance improvements. For example, complaints about some aspect of customer service may have precipitated the need for action whether it be through improving the activity internally or outsourcing it to a supplier.

These issues are explored more fully in the following sections.

7.2   Analysing the type of advantage

7.2.1   Cost analysis

A key element of cost analysis is concerned with comparing the costs of sourcing the activity internally or from an external supplier. Cost analysis in make-or-buy provides a straightforward illustration of the costs associated with carrying out a function internally or sourcing it from an external supplier. The cost elements of

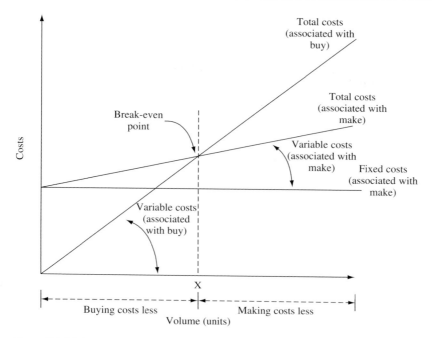

Figure 7.1 Make-or-buy break even analysis

make-or-buy can be analysed using break-even analysis. This is shown in Figure 7.1. If the sourcing organisation decides to make the item there are a number of costs including the fixed costs (FC) of production and the variable costs (VC) associated with producing an extra unit of production. The variable costs vary with demand (D) for the item. In this case the total costs (TC) of 'making' are:

$$TC = FC + (VC \times D)$$

Alternatively, in the case of sourcing the item from a supplier there are no fixed costs. In this case, the total cost of 'buying' is the unit price (P) times the demand (D):

$$TC = P \times D$$

The buying price per unit (P) should include the cost of delivering the item to the manufacturing facility and an assessment of the internal support costs such as purchasing within the sourcing organisation. As shown in Figure 7.1, point (BEP) – or the required demand (D) – is located at the point (X) where the total costs of making and buying are equal. The value of BEP can be simply calculated as follows.

Cost to buy = Cost to make

$$P \times BEP = FC + (VC \times BEP)$$

$$(P \times BEP) - (VC \times BEP) = FC$$

$$BEP = FC/(P - VC)$$

If total demand for the product is less than BEP, then buying from the supplier is the preferred option. Alternatively, if total demand for the product is greater than BEP, then making is the preferred option.

Applying this approach to cost analysis in the make-or-buy decision is relatively straightforward. However, this approach does not account for the fact that there may be constraints upon resources within the sourcing organisation. These factors are sometimes referred to as limiting factors in an accounting context. Limiting factors can include a lack of sufficiently skilled staff, a lack of available funding, a shortage of materials or insufficient facility space. Clearly, the presence of these factors can impact considerably upon the ability of an organisation to perform certain activities internally. For example, in the public sector, funding constraints can inhibit the increased provision of services due to a lack of the necessary infrastructure or skilled staff. However, it is important to have a means of prioritising the use of scarce resources for performing activities internally. This involves using the limited resources in a way that customer demand is satisfied and profit is maximised. These limiting factors can be integrated into the make-or-buy decision to further enhance the analysis. Gordon Ellis (1992) has provided a methodology for analysing the influence of limiting factors in the context of the make-or-buy decision. This approach begins with determining the advantage to the sourcing organisation of the item being produced in-house. In order to do this, the cost of sourcing the item from an external supplier must be compared with the marginal cost of producing the item internally. The next stage involves relating the advantage of making to the limiting factors concerned. Once this is done, it is then possible to rank in order of priority the available resources based upon the advantage to the organisation. This approach is illustrated in the following example.

Table 7.1 shows the cost of buying and the marginal cost of making a number of components required by a manufacturing company for its product portfolio. It is shown that there is an advantage in making all the items internally except for Component 4. In relation to Component 4, there is a disadvantage in producing

Table 7.1 *Cost of buying and marginal cost of making*

Component	C^1	C^2	C^3	C^4
Cost of buying (buy)	£20	£10	£15	£20
Marginal cost of production (make)	£15	£8	£12	£25
Advantage in making	£5	£2	£3	−£5

Table 7.2 *The effect of the limiting factor on internal production*

Component	C^1	C^2	C^3
Cost of buying (Buy)	£20	£10	£15
Marginal cost of production (Make)	£15	£8	£12
Advantage in making	£5	£2	£3
Labour hours per unit	2	4	4
Advantage per hour in making	£2.50	£0.50	£0.75
Ranking	1	3	2

Table 7.3 *Sales demand and total hours required*

Component	Sales demand for end product (units)	Labour hours per unit	Total hours required
C^1	100	2	200
C^2	120	4	480
C^3	150	4	600

the component internally and if appropriate this component should be sourced from an external supplier.

The next stage in the analysis involves relating the limiting factor to the advantage associated with making the component internally. For purposes of illustration, the limiting factor in this example is the availability of labour hours. The effect of this limiting factor on the three components that are suitable for internal production is shown on Table 7.2. It is now possible to express the advantage in making in terms of labour hours and determine the order of priority for allocating the available resources. For example, it can be seen that Component 1 delivers the greatest advantage for the company in terms of labour hours when it is made internally.

Resources can now be allocated based upon the rankings calculated from the limiting factor. Table 7.3 shows the sales demand for the products associated with each component and the total hours required. The total labour hours required is 1280. However, for purposes of illustration, the organisation has only 1000 labour hours available internally and there is a shortfall of 280 hours. Therefore, in this scenario, the organisation will have to buy in some of their requirements to meet end sales demand.

Table 7.4 shows the most appropriate allocation of labour hours to each component based upon the rankings of advantage. It can be seen that the labour hours available should be allocated firstly to both Components 1 and 3 and then the remainder to Component 2. The shortfall for Component 2 should be sourced from an external supplier.

Table 7.4 *Allocation of labour hours based upon rankings of advantage*

Component	C^1	C^2	C^3
Allocation of labour hours available internally	200	200	600
Make (units)	100	50	150
Buy (units)		70	

The benefit of this approach is that the costs associated with sourcing the component externally and producing the component internally can be combined with the influence of the limiting factor. It also illustrates the advantage of sourcing an activity internally in comparison with that of outsourcing. Scarce internal resources can be allocated based upon their priority in terms of achieving a cost advantage for the sourcing organisation.

Identifying cost drivers

These illustrations of cost analysis in make-or-buy are employing conventional cost accounting approaches. Conventional accounting is appropriate when direct labour accounts for most of the total costs and the organisation produces a limited number of standard products that share the same processes. In conventional management accounting, cost is primarily a function of output volume. Cost concepts related to output volume – fixed versus variable cost, absorption costing versus marginal costing, cost–volume–profit analysis, break-even analysis, and contribution margin – have influenced much of the thinking and writing on cost accounting (Shank and Govindarajan, 1993). It is often assumed that direct labour cost varies in line with changes in the volume of output. However, analysis of the practices of companies such as Hewlett Packard, Texas Instruments and General Electric has revealed that the simplistic and rough allocation of indirect costs on the basis of direct labour costs has not worked for many organisations particularly those involved in heavy automation or with short product life cycles or high product complexity (Berliner and Brimson, 1988). Output volume cannot capture all the intricacies associated with cost behaviour. It is widely recognised that the volume approach is too simplistic in an environment with complex processes and products. For many organisations, direct costs now represent a smaller percentage of total costs than support costs. Knowledge workers such as designers and software engineers have displaced much of the direct labour force in many manufacturing facilities. In many cases, overhead costs associated with engineering, marketing and distribution has increased to exceed that of direct labour (Gunasekaran, 1999). The appropriate allocation of indirect costs has now become important for strategic decision-making. The focus on volume tends to understate the cost-per-unit of products that have lower sales volumes and high

degrees of complexity. This problem is further exacerbated in service organisations where the type and complexity of the service influences the cost with volume
having little impact upon the relationship between input and output factors.

Identifying the drivers that influence the costs of organisational activities is a
valuable means of assessing the cost position of the sourcing organisation relative
to suppliers and competitors. The volume driver alone determines little of the richness
of cost behaviour. Cost drivers differ across organisational activities. Analysing costs
from an activity perspective involves identifying the cost drivers that explain the
differences in costs across each organisational activity. For example, the number of
orders received is the cost driver for the receiving activity, number of set-ups is the cost
driver for the production control activity, and the number of orders despatched is the
cost driver for the distribution activity. Cost drivers associated with a number of
organisational activities are shown on Table 7.5. The data requirements for this
analysis of cost drivers are quite formidable. Management can break down the
organisation's functional cost accounting data into the costs of performing specific
activities. The appropriate degree of disaggregation depends upon the economics of
the activities and how valuable it is to develop cross-company comparisons for
narrowly defined activities as opposed to broadly defined activities. At a general
level, there are a number of drivers of costs that impact upon the relative cost position
of the sourcing organisation in relation to suppliers and competitors.

- *Factor costs* – one of the most prominent influences in relation to cost analysis in
  outsourcing evaluation is that of factor costs. Factor costs refer to the inputs
  that are used to perform organisational activities and include labour, capital,

Table 7.5 *Sample activities and cost drivers*

Activities	Cost drivers
Procurement	Location of suppliers
	Relative bargaining power
	Volume of purchases per supplier
Manufacturing	Scale of facilities
	Plant location
	Level of automation
	Labour rates
	Capacity utilisation
Design	Number and frequency of new product introductions per planning period
	Average sales per new product introduction
	Labour rates
	Labour skills availability
Customer service	Number of customers
	Sales per customer
	Service level offered

land and raw materials. For example, external suppliers may have a significant cost advantage in a number of areas as a result of having lower factor costs. Perhaps, the most common influence in relation to factor costs is that of labour rates. Many organisations can realise significant cost reductions by outsourcing to suppliers that have much lower labour costs. As well as considering the labour cost per employee, Bruck (1995) argues that the total annual labour costs should be considered. Total annual labour costs – including ancillary costs such as unemployment insurance – divided by the number of hours worked will yield net labour cost per hour. In many cases, suppliers work longer hours than their counterparts in original equipment manufacturers (OEMs). If this is added to the lower annual labour cost per employee then the personnel cost advantage per hour reaches 30 to 40%. Organisations have realised significant cost savings in areas ranging from manufacturing to telephone call handling by outsourcing to either local suppliers or foreign supply sources.

- *Economies of scale* – can be obtained through internal development or out-sourcing to suppliers. Economies of scale in certain processes can be realised internally through the achievement of high relative market share in the sourcing organisation's respective product markets. Alternatively, suppliers are often in a position to realise scale economies because they are supplying the same products to a number of customers. But what is often not realised is that economies of scale may also exist in material costs, especially if a manufacturer is paying more for all the various components together than it would for a complete outsourced system.
- *Experience* – costs can be reduced through experience built up over time that competitors can find extremely difficult to replicate.
- *Complexity* – the range of products and services offered by an organisation influences costs. Enhancing value for customers through adding unique product or service features is also likely to increase costs. Complexity can dilute the impact of reduced costs associated with other cost drivers such as scale economies. For example, introducing a constant stream of new and innovative products and services can make it extremely difficult to realise scale economies.

Activity-based costing

Interest in cost drivers has been fuelled by the development of activity based costing (ABC). ABC has emerged as a useful guide to management action that can translate directly into higher profits (Cooper and Kaplan, 1991). The ABC approach is broadly applicable across the spectrum of company functions, having important strategic implications. Conventional cost accounting systems focus on units of particular products. Costs are allocated to products because each unit is assumed to consume resources. By contrast, ABC systems focus on the activities performed in the production of products or delivery of services. Costs are traced from activities to products based on each product's consumption of the activities. ABC views the organisation as a collection of activities that incur costs whilst products or services consume activities in their production. The allocation bases, or 'cost drivers', used in ABC are therefore measures of the activities performed. ABC acknowledges that products or services do not directly use up resources: they use up activities. ABC analyses activities in order to provide information that can assist in management decision-making. The primary objective of ABC is to understand the behaviour of costs associated with each activity. There are a number of stages in the ABC approach.

Create a cost pool for each activity

The first stage involves creating cost pools for the activity under consideration. The choice of cost pools is influenced by an identification of the major activities, which create overhead cost. For example, in the case of the procurement activity potential costs pools include the purchase order issue, requisition handling, expediting, payment approval, supplier evaluation, etc. This type of information can be gleaned from analysing the activity concerned.

Allocate overhead costs

The next stage involves allocating overhead costs to each of the cost pools associated with the activity under scrutiny. The objective is to determine how much is being spent by the organisation on each activity. The cost allocation will be determined on the basis of the resources being consumed in relation to each cost pool. For example, referring back to the procurement activity, the supplier evaluation cost pool, overhead cost may be allocated on the basis of labour and equipment involved in undertaking the activity.

Determine the cost drivers

This stage is concerned with determining the underlying factors that cause the activity to occur; i.e. the cost drivers. Cost drivers are related to the number of times an event occurs. The cost drivers are the link between resources and activities. Cost drivers provide an explanation of why costs in activity cost pools change over time (Kennedy, 1996). For example, potential cost drivers for purchase order issue will be the number of orders and the number of suppliers. Cost drivers are likely to be understood by the manager and personnel involved in the activity under scrutiny.

Allocate activity costs to products/services

This stage involves assigning the costs associated with each activity according to the product or service's demand for activities. The cost driver must be in a measurable form that allows it to be identified with individual products or services. The cost drivers are used as a measure for product or service demand. For example, the number of set-up changes (a cost driver) in the manufacturing process combined with the product type (i.e. assuming that product type will influence the number of set-up changes) will determine the overall manufacturing cost for the product.

Proponents of ABC accounting argue that it allows organisations to charge costs more accurately than conventional accounting methods because it allocates overhead far more precisely. For example, consider a bakery that produces two types of baking products – one low volume (*Product A*) and one high volume (*Product B*). A major cost associated with the baking process is switching production from one product to another. The total costs include the cost of the ingredients, direct labour hours and production overhead – which includes the costs associated with resetting equipment to switch production from one product to another. Under conventional cost accounting, the costs associated with resetting production will be allocated equally to both the high and low volume products. The bakery has total production overhead costs of £1 000 000 and total production output of 10 000 000 units. Based upon these figures the overhead rate is calculated as follows:

$$\text{Overhead rate} = \text{total production cost/total production output}$$
$$= £1\,000\,000/£10\,000\,000$$
$$= £0.10 \text{ per product}$$

Table 7.6 shows how the bakery determines production costs for each product using volume as the primary cost driver.

However, this approach fails to accurately identify the costs associated with baking each product. In particular, it underestimates the costs associated with producing the low volume product. Employing the ABC approach provides different results through segmenting the production process into activities. Table 7.7 shows the ABC estimate costs including the cost drivers, the associated cost pool, and the amounts that each cost driver consumes which determines the cost rate.

Table 7.6 *Unit cost using the volume driver*

	Product A	Product B
Total unit production	5000	120 000
Ingredients cost/product	£0.10	£0.05
Direct labour cost per product	£0.10	£0.06
Overhead rate	£0.10	£0.10
Total cost per unit	£0.30	£0.21

Table 7.8 shows the cost drivers and the quantities of each driver used by each product.

The next stage involves calculating the unit cost for each product based upon the ABC data as shown in Table 7.9.

Comparing the ABC unit cost with the unit cost under the conventional accounting approach, it can be seen that there is a major difference in cost for the low volume product. The cost difference is a result of the set-up costs associated with changing production from one product to another. Segmenting the production process into activities allows the costs of resetting production due to changeovers in the production of each product to be identified. Using the ABC

Table 7.7 *ABC costs*

Cost driver	Cost pool		Driver amount		Rate
Set-ups	£150 000	divided by	10 000 set-ups	=	£15
Baking time	£750 000	divided by	250 000 hours	=	£3
Customer orders	£100 000	divided by	10 000 orders	=	£10
Total costs:	£1 000 000				

Table 7.8 *Activity data for each product per annum*

	Product A	Product B
Number of set-ups	200	50
Baking time (hours)	50	200
Number of customer orders	40	30

Table 7.9 *ABC cost per product*

	Product A	Product B
Set-ups (£15 × 200, 50)	£3 000	£750
Baking time (£3 × 50, 200)	£150	£600
Customer orders (£10 × 40, 30)	£400	£300
Ingredients cost	£500	£6000
	(£0.10 × 5000)	(£0.05 × 120 000)
Direct labour cost	£500	£7200
	(£0.10 × 5000)	(£0.06 × 120 000)
Total Cost	£4550	£14 850
Divided by annual volume	5000	120 000
Average cost per product	£0.91	£0.12

approach, the costs of resetting equipment are calculated and charged against the production run for each product regardless of the size of production run. In this case, costs are allocated to each product on the basis of the resources it consumes. Using ABC does not change the total costs. However, costs are assigned differently. Costs are assigned to products based upon the activities they consume rather than volume. The strength of ABC is that it identifies where costs are generated. Under conventional accounting methods, the analysis focused on the administrative or responsibility unit rather than a cross-functional process or activity. ABC evolved around the same time as the use of the value chain concept became prevalent. Both these approaches incorporate the notion that the way in which an organisation serves customers effectively and efficiently is by co-ordinating action in cross-functional activities or processes (Lebas, 1999).

The ABC approach to analysing the costs associated with organisational activities can assist in outsourcing evaluation. In using the ABC approach in outsourcing evaluation, organisations do not have to commit the entire accounting system to ABC. Careful consideration should be given to the amount of resource committed to employing ABC analysis in outsourcing evaluation. The benefits should outweigh the costs of carrying out the analysis. In effect, the level of resource committed should reflect the costs of carrying out the analysis and the level of accuracy required. The principles associated with ABC can be used to inform the analysis of performing activities internally. In organisational capability analysis a key issue to determine is the relative cost position on the various cost drivers of the sourcing organisation in an activity in relation to both suppliers and competitors. Using activity analysis allows consideration to be given to the relationship between costs and price of the product or service and the ways in which value can be added to differentiate products or maximise the price that can be realised. Many organisations have outsourced activities to suppliers with lower labour rates in order to reduce their own allocation of overheads. However, focusing primarily on labour overhead may conceal potential cost increases in other areas of the business. For example, in the case of moving production offshore to countries with lower labour rates, additional costs in shipping and distribution costs will be incurred. Moreover, such a strategy may increase time to market for the introduction of new products particularly if previously designers worked closely with the local internal manufacturing source. Increases in the time between the design of the product and its introduction to the market is likely to create additional cost in general management activities. Conventional accounting does not provide an understanding of what actually drives costs. Employing cost analysis at the activity level allows an organisation to determine more precisely the costs associated with direct labour, direct materials and overheads required to perform an activity.

This analysis can also be related to the overall strategy of the organisation. Referring back to *Chapter 6 – activity importance analysis*, it was shown how the potential sourcing options enabled an organisation to influence both the cost and differentiation dimensions. Clearly, as illustrated in the case of outsourcing activities offshore, analysing costs at the activity level is valuable in the evaluation of the implications of outsourcing for the cost position of the organisation. However, activity analysis is also useful in examining the implications of sourcing options for enhancing the differentiation dimension. Understanding who customers are and what they value plays a crucial role in the management of costs. Referring back to the discussion on the value concept in the previous chapter, it was emphasised that customers do not buy products they buy a set of expected benefits. This bundle of attributes influences the customer when purchasing the product or service. These attributes can be both tangible (such as a CD-Writer on a personal computer or a 24-hour help desk for customers) and intangible (such as product image or company reputation). Each of these attributes incurs costs for the organisation and can be linked back to organisational activities and cost drivers. In the case of attributes that are a potential source of differentiation, the organisation may decide to offer these even though they incur extra costs. For example, an organisation may decide to offer a higher level of service in the form of a 24-hour call-out service in the event of product breakdowns in order to differentiate itself from competitors. This service may be provided internally or through third-party service providers. Organisations can make rough estimates of the costs of adding differentiating attributes to the total product and service offering. ABC also provides useful insights into what is being done with the resources allocated to perform an activity. In particular, it can provide a sound indication of the value of an activity to the organisation. For example, activities that consume a considerable amount of resource yet add little value in the eyes of the customer can be identified. This resource can include both management time as well as direct labour. These activities may be considered as potential targets for outsourcing in order to redirect employees and management to more value-adding activities. In addition, by assessing the costs associated with an activity the sourcing organisation can make an assessment of its cost base in relation to both competitors and suppliers and identify opportunities for cost reduction. For example, an activity can be changed or simplified in order to reduce costs and enhance performance. Benchmarking can serve as a valuable tool in supporting this analysis and in particular, in the comparison of competitors and suppliers and the identification of potential areas for improvement.

Illustration 7.1

Activity-based costing at Wavin
Wavin applied some of the concepts associated with ABC to its manufacturing operations. Wavin is a large manufacturer of plastic pipe systems in Europe. It

manufactures and distributes a wide range of products that extends into markets ranging from domestic pipe systems to gas, water gutter, drainage, sewer, cable ducting and irrigation. As part of its manufacturing excellence programme, it segmented its products into fast-moving and slow-moving in order to adopt a more focused approach to production and logistics cost analysis. Previously, all its products were treated on the same basis for costing purposes. For example, overhead costs were allocated to products on the basis of machine size, efficiency and utilisation as well as product family, colour and type. However, the segmentation of its product portfolio into fast-moving and slow-moving required a new approach to costing because the manufacturing and logistics approach were different for each of the product classifications. Using activity analysis to analyse each group of products, the company found that the fast-moving products managed themselves in their own manufacturing cells, whilst the slow-moving products demand a significant level of management attention from material supply to customer demand. There were two significant benefits of adopting this approach.

- Understanding the relationship between product costs and manufacturing resources allowed the company to identify opportunities for managing the supply of resources, which had previously been treated as overheads in the manufacturing process.
- It allowed the company to identify the overall profitability of its product portfolio, information that could be used in business strategy formulation.

Wavin experienced some early benefits from applying this approach to certain parts of its business. For example, Wavin applied this approach to one of its market segments that it considered unprofitable. The resulting information provided a guide for strategic action that allowed the company to increase profitability in this market segment. Although there were some difficulties with allocating costs accurately to individual products, considerable emphasis was placed on achieving accuracy of activity-based costs allocated to product families, the level at which decisions were made. It also adopted a more focused approach to the way in which it managed relationships with its customers. Important customers were classified on the basis of the following characteristics:

- the volume of business was at the level where economies of scale could be achieved;
- some costs were directly attributable; and
- the customers were regarded as strategically important regardless of their turnover.

Although only a small number of its customers matched this profile, they represented a significant percentage of the turnover of Wavin. This approach also allowed Wavin to allocate costs more accurately to these larger customers. Using the activity-based approach allowed Wavin to segment the organisation into activities that could be used as a basis for pooling costs. Analysis could also be carried out of how finished products were routed through different processes depending upon their type and complexity. Therefore, costs could be allocated on a more focused and differentiated way and activities could be measured and targeted for improvement. Moreover, the existence of certain activities and their associated costs could be questioned.

An example of how it used this approach in analysing distribution costs provides an interesting illustration of the potential benefits. The company used a combination of own-fleet and subcontracted vehicles to manage the distribution of its products. A major issue for the company was to manage effectively the utilisation of its own fleet of vehicles. In order to do this, it used the principles associated with ABC to analyse the relationship between sales and transport costs. The analysis revealed the following relationships:

- *activity centre* – including own-fleet transport and subcontracted transport;
- *cost drivers* – product types, product families, customers and regions; and
- *performance measures* – volume per load for product mix, cost per load and cost per mile.

Using this approach Wavin found that different product families and types influenced costs in different ways. For example, the volume of rainwater guttering that could be delivered on a single lorry was far greater than the volume of down-pipe. The importance of effective transport utilisation for a company delivering piping systems is illustrated by the fact that the distribution of pipes is like shipping air. By differentiating between products the company could optimise the mix for each delivery.

Source: Marshall, B. (1995). Activity-based costing at Wavin, *Management Accounting*, May, 28–30.

7.2.2   Benchmarking

The importance of cost performance in capability analysis has already been illustrated. However, cost performance is only one aspect of capability analysis. As well as considering costs, the relative performance along a number of other dimensions such as quality, flexibility and service has to be considered. Analysis along these additional dimensions is important given the experiences of organisations that have embarked upon outsourcing primarily on the basis of costs. For example, initial analysis may reveal that a supplier possesses a lower cost position whilst at the same time being weaker in areas such as quality and service. Organisations must have an understanding of the key measures that indicate performance in an activity relative to suppliers or competitors. Benchmarking can assist with this analysis. The Japanese word *dantotsu* means striving to be the best of the best. It captures the philosophy of benchmarking. Benchmarking is a proactive process of changing operations in a structured way to achieve superior performance over competitors (Sweeney, 1994). The American Productivity and Quality Centre (1993) provides a comprehensive definition of benchmarking as the process of:

- comparing practices and results with the best organisations in the world, and adapting the key features of those practices to one's own organisation;
- accelerating organisational learning, customer-driven quality and continuous improvement;
- helping organisations identify breakthroughs, by comparing their processes to those of the organisation that is recognised as the best; and
- helping organisations learn from each other whether it be in business, health care, government, or education.

Therefore, benchmarking is a continuous process of measuring product, service and processes against those organisations that are recognised as competitors or world leaders in their areas. The purpose of such comparisons is to enable companies to determine where and how performance can be improved. Benchmarking is concerned with searching for and implementing best practice and performance improvement. However, this does not mean imitating processes used by other companies. Rather, it is concerned with understanding why another organisation carries out activities more effectively than its peers do. Such an approach will then provide companies with a basis to determine how performance can be improved. The objective of benchmarking is to understand and incorporate process and product innovations that have been successful in other contexts. Benchmarking is not purely output-driven, but on the process employed to achieve that output. It is argued that organisations that are characterised as being dynamic in their learning and change processes will be more successful than organisations that are not (Senge, 1990). Benchmarking should provide a systematic process to learn and adapt an organisation to best practice.

Benchmarking can highlight those activities and internal processes in which the organisation has superior performance or cost advantage in relation to competitors. In fact, companies are realising the benefits of combining the benchmarking of value chain activities with business strategy. It can identify potential sources of competitive advantage that can be exploited more fully to develop capabilities that are difficult to replicate. It also assists in identifying weaknesses that need to be addressed to become more competitive. The advantage of using benchmarking is that it aligns the operational activities at the lower level with the overall business strategy of the organisation. It is also important to note that the needs of the customers of the organisation can be linked to the benchmarking process. Benchmarking assists organisations in optimising their capability to meet the needs of customers by ensuring that processes are more superior, consistent and effective than competitors. Therefore, it can be used as a basis for building competitive advantage.

Benchmarking complements the use of traditional financial measures to assess business performance. For example, concentrating exclusively on financial indicators such as 'return of capital employed' (ROCE) and net profit margin can lead to limited attention being paid to other critical factors such as research and development (R&D) capability and customer satisfaction. Financial measures are compiled historically and can become irrelevant by the time decisions have to be made. Furthermore, financial measures can lack relevance to the processes that are valued by the end customer. Therefore, there is a need to supplement traditional financial performance measures with non-financial indicators of the processes that contribute to their achievement. Kaplan and Norton (1992) have argued that non-financial measures should be given the same level of emphasis when determining strategy, promotions and the allocation of resources. Eccles and Nohria (1992) have argued that there are a number of compelling reasons for using non-financial performance systems such as benchmarking.

- The use of non-financial measures encourages management to adopt a broad range and long-term view of organisational performance.
- Measures of non-financial variables target events that are meaningful and actionable in managerial experience; e.g. delivery times, market position, and employee turnover.
- Non-financial measures can serve as leading indicators of financial results and thus be of value to those for whom profit is the principal measure of business performance.

Therefore, a critical part of benchmarking is the selection and manipulation of metrics to represent the performance of the activity under analysis. Table 7.10

Table 7.10 *Sample productivity measures (Hiltrop and Despres, 1994)*

Function	General measure	Specific measures
Engineering	Engineering per facility	Process engineers per product line
Finance	Revenue per finance employee	Revenue per accounts payable employee
Marketing	Revenue per marketing employee	Marketing expenditures per marketing employee
Quality	Revenue per quality employee	Proportion of quality employees devoted solely to quality activities
Research and development	Annual new product introductions per R&D employee	Process improvements per R&D employee

From Elsevier. Reproduced with permission.

shows a sample of potential performance metrics for use in the measurement of the productivity of a business function.

Benchmarking can be classified as follows (Carpinetti and de Melo, 2002).

- *Process benchmarking* – is used to compare operations, work practices and business processes. Process benchmarking involves developing a detailed understanding of how a particular process is carried out and comparing it to how it is carried out and measuring it against performance levels in that process with other organisations.
- *Product benchmarking* – is used to compare products and/or services. This is similar to reverse product engineering that focuses on the analysis of specific components and functions of the products of competitors. Reverse engineering can also act as starting point for process benchmarking. For example, the analysis can be extended to include an examination of the processes that underpin the products of competitors.
- *Strategic benchmarking* – is used to compare organisational structures and management practices and business strategies. Organisations may have similar approaches or initiatives to achieve strategic objectives that are comparable.

Potential benchmarking partners may include other functions or sites within the same organisation, current or potential suppliers, competitors in the same or different geographical territories and organisations in related or unrelated industries. It is quite common that companies benchmark companies in other industries. For example, Southwest Airlines had a major problem with most of their aircraft being on the ground between flights for an average of 40 minutes. Benchmarking its refuelling processes against other airlines revealed that it was already one of the industry leaders. However, by looking outside the industry for the most efficient re-fuellers in the world – Formula One Racing – it adopted their turnaround processes used during pit stops to reduce the refuelling time to 12 minutes (Murdoch, 1997). It is important to consider carefully which partners to use in the benchmarking process. Drew (1997) has identified the benefits and limitations associated with the different types of partners as shown in Table 7.11.

Integrating benchmarking into outsourcing evaluation

In order to integrate benchmarking into outsourcing evaluation it is important to understand and prioritise the activities under scrutiny. It is not feasible to undertake an extensive benchmarking exercise for each activity under scrutiny for outsourcing purposes. In fact, the evaluation of outsourcing for an organisation is likely to focus on one or a limited number of activities as a starting point rather than an evaluation of every organisational activity. The prior experience of the organisation in the area of benchmarking is a key consideration. If it is the organisation's first experience with benchmarking then it may be beneficial to undertake a brief and highly focused exercise that promises to deliver a high impact. Also, the type of activity will have a significant influence on the way in

Table 7.11 *Benefits and limitations of benchmark partners (Drew, 1997)*

Type of partner	Benefits	Limitations
Internal	Breaking down internal barriers; improved communication and information sharing; ease of access to partner cost.	Does not identify global 'best practices'; internal politics/'turf guarding'.
Competitor	As part of consortium can share costs and effort; ease of identifying partners; opening up to ideas from outside.	Legal, proprietary issues.
Related industry	Study of 'best practices' in generic functional or business process management; as with competitors (above).	Legal, proprietary issues: harder to identify partners.
Unrelated industry	As for a related industry, but you may have a greater chance of discovering new ideas and breakthroughs.	Harder to identify appropriate partners.
International	Identifies global 'best practices'; as with unrelated industry (above).	Cost, time, effort; difficulty of identifying partners.

From Elsevier. Reproduced with permission.

which outsourcing evaluation can proceed in terms of capability analysis as outlined in the following examples.

- *Non-critical* – the sourcing organisation may approach a supplier with which it is currently doing business in order to determine if the supplier is interested in taking on an increased level of business in a particular product area. The supplier may already have the necessary capability in this area. This increased business may be in the form of a portfolio of components that the organisation wishes to completely outsource and close down the internal production source. Alternatively, the organisation may decide to approach a supplier that it is not familiar with.
- *Critical* – in the case of a more critical activity, the sourcing organisation may have to undertake a rigorous analysis of potential external sources to determine performance levels.

Illustration 7.2

Benchmarking classifications based upon partner type
Classifications of benchmarking can also be based on the type of partner, as follows (Camp, 1989).

- *Internal benchmarking* – by comparison of performance of units or departments within one organisation. Even though not explicit in this definition, comparison can also be made of similar products or services of similar business units. Internal benchmarking is particularly valuable for multi-site organisations where similar activities are carried out at the different sites. Similar activities are normally found in other parts of the same organisation. For example, dealing with customer queries is a process that can be found not just in the one site but also across a number of sites that deal directly with customers. Therefore, it is useful in determining how the activity is carried out within the same organisation. A major advantage of this type of approach is that full access can normally be obtained to understanding how the process is carried out. Good practices and innovative approaches to activities can also be readily shared and transferred across sites. However, a limitation of internal benchmarking is that it is likely that most parts of the same organisation will be carrying out similar processes in much the same way. Therefore, this may limit the potential to learn and create radically improved processes.
- *Competitive benchmarking* – by comparison of performance with competitors. In this case, comparison can be made of products or services and business processes. If carried out effectively, competitive benchmarking can be particularly useful in discovering new and innovative approaches that can enhance service delivery and performance. However, a major limitation is gaining access to competitor information whether it be data on performance information on their processes. Performance levels can also be compared with other organisations that are not direct competitors. For example, customers will compare performance in areas such as call handling response times against other organisations even though they are not direct competitors. In this case, it is important to identify organisations that have high performance levels in the areas that are valued by customers. Once identified, it is then necessary to analyse how these activities are carried out and the performance levels achieved. The advantage of this approach is that benchmarking can lead to changes that can be implemented which directly contribute to customer satisfaction.
- *Functional benchmarking* – specific function comparison with best practice. It is an application of process benchmarking that compares a particular business function in two or more organisations. It involves comparing the structure and performance of a function in an organisation with comparable functions elsewhere. For example, the entire human resource function could be benchmarked rather than one of its key processes such as recruitment and training.
- *Generic benchmarking* – the search for the best practice irrespective of industry. It is similar to functional benchmarking but the aim is to compare with

Table 7.12 *Four types of benchmarking for rail operator (Tomlinson, 1998)*

Type	Process	Examples
Internal	Train hygiene	• Between inter-city routes • With regional express routes • With regional rail services
Competitive	Public transportation hygiene	• Motorway service stations • Airports • Coach stations
Functional	High usage facility hygiene	• Fast food chains • Office/factory services
Generic	High usage, high standard facility hygiene	• Milking parlours • Food factories

the best in class without regard to industry. It is also important to understand who customers of the organisation regard as best-in-class. For example, Xerox benchmarked against the following companies:
• Walt Disney – staff motivation and training;
• American-Express – accuracy of invoicing and billing;
• Marriott Hotels – dealing with customer complaints.

Table 7.12 illustrates how a rail operator could use these four types of benchmarking.

Sources: Camp, R. C. (1989). Benchmarking: the search for industry best practices that lead to superior performance. New York: *ASQC Quality Press*. WI.Tomlinson, G. (1998). Comparative analysis: benchmarking. In: *Exploring Techniques of Analysis and Evaluation in Strategic Management* ed. V. Ambrosini, G. Johnson and K. Scholes. Hemel Hempstead: Prentice Hall Europe, pp. 62–78. © Prentice Hall. Reproduced with permission.

Each of these types of activities has different implications for the depth of analysis that should be undertaken. For example, in the case of a critical activity, the capability analysis may involve the following comparisons:
• The sourcing organisation may compare its own performance in the activity in terms of cost, quality and delivery against the supplier. This analysis may also extend to other suppliers in the supply market.
• As well as considering the performance of suppliers, the sourcing organisation may compare its own performance in the activity against numerical performance indicators established and reported in public sources.
• Along with external analysis, the organisation may undertake some historical comparisons in order to determine any significant changes in performance levels.

This type of analysis for less critical activities is not likely to be as in-depth as for critical activities. For example, the analysis may only involve a benchmarking exercise to reveal any significant differences in performance with that of potential external sources. In this case, it may be more prudent to consider allowing an external source to perform the activity rather than expending considerable resource on an expensive benchmarking exercise. However, for critical activities a more comprehensive level of analysis is required due to the potential implications for competitive advantage. This next section outlines a number of the critical aspects involved in undertaking a comprehensive benchmarking exercise for outsourcing evaluation purposes.

Benchmarking approach

This involves deciding who should undertake the benchmarking exercise. The sourcing organisation can either use internal staff alone or involve external consultants to assist in undertaking the benchmarking exercise. A team should be formed to carry out the benchmarking of the activities under scrutiny. This team will be composed of both personnel involved in outsourcing evaluation, such as senior management, as well as representatives from the activity under scrutiny. In order to ensure objectivity in the analysis, leadership of the team may be entrusted to a senior manager rather than the functional manager of the relevant activity under scrutiny. It is crucial to involve those who are directly involved in the activity to help create ownership of the exercise. Where internal staff alone are undertaking the analysis, training and support for those involved in benchmarking will be required. However, in many instances organisations will involve third parties such as external consultants or benchmarking service firms. This approach is particularly appropriate in the case of important organisational activities. Although this may increase the costs of the exercise there are a number of potential benefits including instant expertise and experience; access to other benchmarking data; contact with potential benchmarking partners; and additional credibility as independents (Wisniewski, 2001). Consultants can assist in developing process maps of the activity under scrutiny, and define the project, its task and scope. Consultants can also assist with educating the team about the various tools and techniques available. Staff affected by the benchmarking exercise should be made aware of its purposes. Staff are likely to adopt a defensive attitude if the purpose of the exercise is to confirm an outsourcing decision that has already been made. Therefore, staff should be informed that benchmarking is only one part of the evaluation. Benchmarking should be employed to ensure that the sourcing option chosen supports the business strategy of the organisation whether that is continuing to perform the activity internally or outsourcing the activity. It should be regarded as an opportunity for staff to demonstrate their capabilities. Staff should also be given the opportunity to identify any areas for potential improvements if there are any deficiencies in performance. Therefore, communication is required at all stages of the benchmarking exercise for those involved and those affected. Without communication, staff will perceive it as just another cost-cutting programme that is being imposed from senior management.

Analyse the activity

This is concerned with understanding and analysing the activity under scrutiny. The level of analysis will significantly depend upon the importance of the activity. In the case of an activity considered to have a low level of importance it may only be necessary to carry out a data benchmarking exercise that involves numerical comparisons of performance against published performance indicators. However, in the case of an activity deemed to be of critical importance it is necessary to conduct more extensive analysis to determine the strategic options in terms of either improvement or using an external source. This more extensive analysis should involve carrying out the following tasks:

- identification of the people and their roles in the activity;
- determining the role of suppliers in the activity; and
- mapping and documenting the activity by interviewing the relevant people involved in the activity and analysing organisation documents and reports.

Process maps are detailed diagrams comprising a flowchart of tasks and activities that make up some process. They also contain information for each step in the process including key input requirements; resource requirements; critical controls and constraints on the process; required outputs; performance levels or standards; and customer requirements (Foot, 1998). Obtaining co-operation from the staff involved in the process can significantly enhance the validity of the process map. Process maps can also be used to compare performance across organisations to identify areas for improvement.

Performance metrics should also be developed. Before comparing performance with external sources, a clear understanding of internal performance is required. This is quite a challenging task in the benchmarking exercise. Although many organisations believe that reliable measures of performance are used to manage their operations effectively, many still lack the appropriate measures to achieve superior performance. The choice of key metrics to be used for benchmarking purposes must be based upon how well they measure process performance, which in turn contribute to the overall success of the organisation. When searching for metrics it may be prudent to select metrics that are already in use in the organisation. These metrics are already available and recognised. The metrics chosen can be related to the critical success factors or drivers of business success. These relationships can be mapped on to a table as shown in Table 7.13, which illustrates an example from a research and development company.

The activity or process is shown in the first column. In the second column, the critical success factor is identified. Finally, the third column contains the metrics

Table 7.13 *Focus-performance-measure illustration for an R&D company (Karlof and Ostblom, 1995)*

Focus	Performance	Units of measurement
Project work in an R&D company	Specifications of Requirements	Customer perceived Quality Costs Lead times
	Project planning	Costs Time Number of revisions
	Goal Management Deviation Component synergy Project budgeting	Deviation from checkpoints Compliance with specifications Frequency of re-use Cost accounting Follow up

© John Wiley. Reproduced with permission.

that are reliable indicators of the performance of the activity or process. Where possible, quantitative metrics should be used to facilitate comparisons. Qualitative metrics may also be used. However, it must be borne in mind that qualitative metrics can be more difficult to interpret and apply for comparison purposes. A sample of metrics for a number of performance areas is shown in Table 7.14.

Potential partner identification

Once the processes have been mapped and understood, suitable benchmark partners should be found. Based on the mapped processes and the metrics developed, the development of a partner profile can assist in identifying companies carrying out similar processes. The choice of benchmarking partners will be influenced by the following factors (Wisniewski, 2001).

- The level of detail at which benchmarking is carried out.
- The extent to which potential partners have developed their own process maps and a detailed knowledge of their own performance.
- The extent to which potential partners are willing to share detailed information on performance and processes.
- Costs involved in working with particular partners in other geographical areas.

Ideally, the organisation should benchmark against partners that are best-in-class in the process under scrutiny whether they are competitors, suppliers or organisations in other industries. However, the benchmarking partner chosen may be influenced more by the willingness of the partner to become involved than on their level of competence on the process under scrutiny. For example, this stage may be driven primarily by the ease of access that can be obtained from the

Table 7.14 *Sample metrics for performance areas (Karlof and Ostblom, 1995)*

Performance criteria	Metric
Market share	Units
	Revenue
Profitability	Margin contribution
	Return on capital employed (ROCE)
Competitors' growth	Market share per segment
Materials	Proportion of total cost
	Price/volume
	Distribution costs
Direct/indirect human resource costs	Proportion of total cost
	Number of employees per function
	Productive hours per employee
Capital costs	Rate of turnover:
	• Total assets;
	• Fixed assets;
	• Inventory.
	Leasing costs
Service	Response time
	Average time of service call
	Order processing routines
	Production planning

© John Wiley. Reproduced with permission.

potential benchmarking partner. Karlof and Ostblom (1995) have identified a number of sources for potential partners.

- *Trade associations* – promote the collective interests of organisations within their industries. These associations normally have information at both an industry and organisation level.
- *Databases* – there are now commercial databases available, which provide information on industries and their respective organisations. Many of these databases are now available on the Internet. For example, Mintel.com provides information on companies across a range of industries.
- *Best practice databases* – there are a number of 'best practice' databases that have been established for benchmarking purposes. Some of these databases collect information on a large number of companies regarded as outstanding in their respective fields. These can be valuable for identifying areas for comparison and potential partners. A number of these databases are now accessible online; e.g. www.benchmarking.co.uk and www.benchnet.com.
- *Published reports and statistics* – there are also many statistics and reports that can identify outstanding companies. Some reports provide information that can be used for comparative purposes.

- *Company networks and organisations* – internal personnel within an organisation can be a valuable source of information. For example, marketing personnel
  tend to be very well informed about industry trends and examples of excellence.
  In fact, internal personnel can be a very useful source for identifying a potential
  benchmarking partner.

In the context of outsourcing evaluation, the most competent supplier of the
activity is a valuable benchmarking partner. Information from suppliers can be
much more readily obtained than from competitors who are likely to have a
natural reluctance to share commercially sensitive information. Futhermore, in
the context of outsourcing evaluation, there are a number of potential advantages
for using potential suppliers.

- It is more straightforward to obtain access to potential suppliers provided that it
  is demonstrated beforehand that the exercise is mutually beneficial. It is important to verify any information provided by a supplier particularly if the supplier
  is likely to win business as a result of demonstrating high levels of performance
  in the activity under scrutiny. For example, suppliers may provide inaccurate
  cost information if there is a potential opportunity of winning business from the
  sourcing organisation. However, a potential advantage of using suppliers as
  partners is that contacting their customers can be another valuable means of
  assessing their performance.
- In the case of a supplier that the sourcing organisation is currently doing
  business with, a great deal of knowledge and information is already available
  on both current and historical performance – such as quality, cost, delivery, etc.
  In fact, this type of performance information can be more valuable than the
  analysis of the activity under scrutiny. However, the rigour of the analysis of the
  process under scrutiny should not be compromised because the sourcing organisation is already familiar with supplier operations.
- If the sourcing organisation is less capable than external suppliers then, there is
  the potential for outsourcing the activity to the more competent supplier
  benchmarked.

In relation to competitors, benchmarking can act as an imperative for strategic
action. For example, if the organisation establishes the performance benchmark
for its competitors then this can serve as a reliable indicator of competitive
advantage. Benchmarking against competitors may also reveal practices that are
not worth attempting to replicate or surpass. Finding that competitors are
more competent then the sourcing organisation is confronted with the following
options.

- In the case of an activity deemed to have a high level of importance, then the
  organisation may attempt to develop the capability and improve to match or
  surpass competitor performance. Alternatively, the organisation may decide to
  select a supplier to perform the activity.

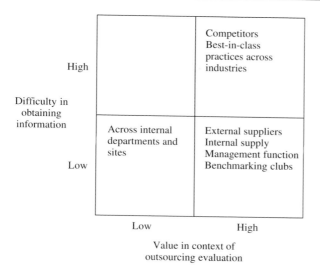

Figure 7.2 The difficulty of obtaining information and its impact upon outsourcing evaluation

- In the case of an activity deemed to have a low level of importance, then it may be more prudent for the organisation to consider outsourcing the activity.

In relation to the use of best-in-class partners across other industries, access may be more readily achievable than that from direct competitors. With full access obtained then it may be possible to determine whether there is the potential to improve performance rather than considering the outsourcing option. Alternatively, the exercise may reveal too significant a disparity in performance that may force the sourcing organisation to consider outsourcing to a more competent supplier. Figure 7.2 illustrates the relationship between the resource involved in acquiring information and relevance to outsourcing evaluation.

Data collection and analysis

Effective benchmarking involves ensuring that the right information is collected and then analysed. The first phase involves designing a questionnaire based on the partner profile. The benchmarking questionnaire should be a clear, formal agreement and understanding about what data and information is to be collected and shared. The questionnaire can be used to establish similarities in operations between the organisation and the partners identified. The questionnaire may be developed jointly between the partners.

In relation to the level of access to and use of confidential information, rules need to be agreed by the partners. As well as a questionnaire, internal reports, site visits, interviews can be carried out with key staff within the partner organisation. Prior to the data collection phase, consideration should also be given as to how the information will be analysed. Data collection and analysis should be carried out in a rigorous manner to maximise the benefits from the exercise. Walleck *et al.* (1991) provide four interesting characteristics of the benchmarking exercise to distinguish it from company visits or plant tours.

- Benchmarking is preceded by both in-depth industry and company analysis, so that the benchmarking team knows what to look for. Due to the variety of activities under analysis, this team should be cross-functional consisting of executives, managers, and specialists in each activity.
- Benchmarking partners should be selected by careful analysis to facilitate valid comparisons. Searching industry databases and interviews with customers, suppliers, financial analysts, and consultants can also assist in providing partner companies for benchmarking.
- Agreeing on measurement formats, integrating data and measures from available sources, and dividing the observation tasks for company visits is an essential part a comprehensive on-site investigation.
- Collecting and interpreting the results is not only the most interesting but also the most difficult task of all. This process should be undertaken with a high degree of rigour which contrasts with plant tours or company visits.

In fact, the success of the benchmarking exercise often depends upon the organisation understanding the results and consequences. In the context of outsourcing evaluation, the outcome of the exercise is likely to lead to significant change in the management of the organisation both internally and externally into the supply chain.

Performance gap analysis

The benchmarking exercise should yield the one of the following three scenarios:

- *parity* – no significant difference in performance between the sourcing organisation and competitors or suppliers;
- *internal more competent* – in this scenario internal performance is superior to that of either competitors or suppliers;
- *external more competent* – the performance of either competitors or suppliers is superior to that of the sourcing organisation.

Effective analysis will enable the organisation to determine how its performance compares with external sources. Essentially, the organisation has to identify the performance gap revealed by the performance analysis. It is important to identify and understand the underpinning performance indicators that explain the existence of the gap. Benchmarking can assist in identifying the potential strategic sourcing options that will lead to improvement in the performance of the activity under scrutiny. Employing benchmarking may lead the organisation to pursue different sourcing strategies relative to competitors. For example, a manufacturing company may discover through benchmarking that its cost position for one of its manufacturing processes is on a par with the industry average. Therefore, the company may decide to outsource this manufacturing process to a supplier with a lower cost base. Alternatively, a benchmarking exercise undertaken by a competitor of the manufacturing company may reveal that it possesses the lowest cost base for the manufacturing process in the industry. In this case, this company will elect to continue to carry out the manufacturing process internally.

It is possible to align benchmarking with the dimensions of performance that are critical to competitive advantage. It can be seen that there is a clear link between benchmarking and the overall business strategy of the organisation. As well as alerting management to potential outsourcing opportunities, it can also act as a vehicle to understanding whether activities can be improved internally. However, it is important to be aware of its potential limitations. Although benchmarking can help an organisation to become more efficient and competitive, it does not create competitive advantage for organisations. Drew (1997) has identified a number of caveats associated with the use of benchmarking.

- Benchmarking is not, in itself, a strategy for achieving competitive advantage. It is a related set of activities that can support and enhance strategies for improvement that may lead to competitive advantage.
- Benchmarking is seen as more than a means of gathering competitive intelligence. Benchmarking is often associated with change management. However, as is the case with many management techniques success is determined at the implementation level where improvement has to be achieved.
- Although the maxim 'adopt, adapt and improve' is often associated with benchmarking, it is not a technique for creativity or innovation. Normally, when breakthrough innovations occur, benchmarking is only one tool used by management. Benchmarking is primarily a technique for developing and implementing strategies that are either imitative or incremental innovations. However, it can also act as a supplement to the process of discovering more radical innovations.

These caveats are particularly pertinent to the evaluation of activities for outsourcing purposes. The importance of understanding the capabilities of competitors in activities that make a significant contribution to competitive advantage has already been emphasised. However, in some cases it is not possible to benchmark or understand activities that are crucial to sustainable competitive advantage. For example, Hamel and Prahalad (1994) have cited Sony's strengths in miniaturisation as an illustration of a core competence. Due to the complexities and knowledge-based resources associated with this core competence, it would be impossible to undertake a valid benchmarking exercise. Futhermore, an organisation is unlikely to be willing to share sensitive information regarding their core competencies lest it would compromise its competitive position. This analysis may lead to the conclusion that benchmarking is only appropriate for peripheral activities. However, it must be emphasised that a core competence can be strengthened and developed by analysing the activities that contribute to the core competence. It is wrong to assume that every activity associated with a core competence is controlled and kept in-house. For example, if one considers an often-cited exemplar of a core competence – Honda's world leadership in engine design and manufacture – it is likely that a number of activities associated with this core competence are being carried out both internally and externally. Honda performs activities for which it has more competent internal sources that in turn support the development of the core competence. It can access suppliers that are more capable at performing an activity that will contribute to the development of this core competence. Benchmarking can be used to determine the most competent provider – either internal or external – of an activity and hence raise standards in the activities that contribute to its overall core competence of engine design and manufacture. By outsourcing those activities where it has no performance advantage and is much weaker than potential external sources it will more readily achieve excellence in this core competence by using that source. It is also possible to use this analogy in the context of the critical success factor (CSF) methodology. For example, an organisation may establish excellence in customer service as a CSF. In order to measure the performance in customer service, the organisation may define the following metrics:

- sales per customer;
- number of complaints;
- number of new customers per planning period;
- number of inquiries converted into customers;
- customer service call waiting time; and
- percentage on-time delivery.

Figure 7.3 illustrates the activities that contribute towards the achievement of excellence in customer service.

Again, analysing the capability of the sourcing organisation in these activities that contribute to customer service may reveal either positive or negative performance levels in relation to potential external sources. In the case of negative performance levels, the sourcing organisation has the option of enhancing performance through either outsourcing to an external source or an internal improvement initiative. However, the crux of the argument is that it is possible to use benchmarking at the activity level to determine whether internal or external sourcing arrangements should be employed to enhance performance rather than at the CSF or core competence level.

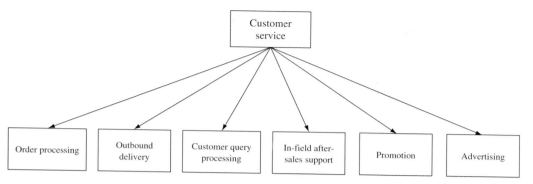

Figure 7.3 Customer service activities

Illustration 7.3

The role of benchmarking in information systems outsourcing evaluation
Based on their analysis of the practices of a number of companies in outsourcing evaluation for information systems (IS) services, Mary Lacity and
Rudy Hirschheim provide a number of valuable insights and lessons on the role
of benchmarking that are relevant to outsourcing in other contexts. The participants in the study reported mixed reactions to the value of benchmarking. On
a positive note, some companies reported how benchmarking had assisted in
identifying areas for cost reduction within the IS function. It was also particularly valuable in convincing senior management that costs could be reduced
internally without having to outsource part of the function. Alternatively, some
companies found that the internal IS function was not competitive in relation to
external vendors even though benchmarking had indicated that cost and performance levels were excellent. Indeed, some IS managers were found to be
employing benchmarking for political motives, using it to show that they were
as competent as any other potential service provider. However, senior management in some cases were sceptical of the results of the benchmarking exercise as
the IS managers had undertaken the benchmarking themselves or hired the
external consultants. Senior management also had difficulties in relating technical measures employed in the analysis with the general needs of the business.
The study undertaken by the investigators revealed some interesting insights
into the role of benchmarking in outsourcing evaluation. In particular, the
study revealed that companies used the following as sources to compare their
performance with external companies.

- *Informal peer comparisons* – this involves using a network of peers to informally assess performance. There was considerable emphasis placed on informal peer comparisons because managers knew and trusted their peers.
- *Informal outsourcing queries* – this involves the manager contacting outsourcing vendors in order to test the market. However, this was found to be of little value due to the lack of response and attention given by outsourcing vendors.
- *Outsourcing evaluations* – this was found to be the most rigorous approach involving creating a request for proposal, requesting vendor bids, and comparing bids against internal costs. Although outsourcing evaluations can be extremely resource intensive, they can provide valuable comparative information.
- *Benchmarking services* – using benchmarking services was found to be a more objective means of assessing performance than informal comparisons. Benchmarking services provided a wide range of comparative data and were perceived as less disruptive to operations than formal outsourcing evaluations.

Based upon the analysis carried out, a number of recommendations were made in relation to both the benchmarks used and the selection of a firm to assist in benchmarking which included the following items.

- *Involve senior management in the selection process* – senior management are more likely to be convinced of the validity and results of the benchmarking exercise if they are involved in the selection of the benchmarking firm.
- *Measure what is important to senior management* – there is a tendency for many IS benchmarking exercises to employ benchmarks that are too technical. Moreover, the benchmarking exercise may focus on performance values that are not important to senior management.
- *Select the most competitive company to benchmark* – rather than select a large reference group, it is much more valuable to compare performance against the 'best of breed' in the particular area being benchmarked. It was found that senior management were much more impressed by the exercise when the 'best of breed' in the area was used as the reference.
- *Gather data during peak times* – measures of performance taken at peak times give a better indication of effectiveness and are more valuable to senior management. The data gathered should also be validated through consulting with the relevant people within the organisation.
- *Repeat benchmarks* – it was found that senior management placed more emphasis on the results of benchmarking if the measures were used and repeated on a periodic basis. This should assist in improving performance over time.

Benchmarking in the context of outsourcing evaluation can be an extremely valuable means of demonstrating to senior management that the IS function is attempting to be efficient and is seeking ways of achieving performance improvements. However, it is important that benchmarking is used to support the overall needs of the business than for example, political motives. The ideal outcome from the benchmarking exercise is creating an environment that allows the IS function to learn new and better practices.

Source: Lacity, M. C. and Hirschheim, R. (1995). The role of benchmarking services in insourcing decisions. In: *Beyond the Information Systems Outsourcing Bandwagon: The Insourcing Response.* Chichester: Wiley.

7.3   Analysing the source of advantage

This stage is concerned with understanding the source of the superior performance in an activity – possessed either by the sourcing organisation or external source. In many cases the causes of superior performance will be obvious. For example, labour costs are likely to be the most important influence on activities that are highly labour-intensive. However, in other situations the causes of poor performance may not be clear. For example, poor quality in an activity may be as a result of a combination of factors including poor management and a lack of employee support. Many organisations have embarked upon outsourcing upon discovering that performance is lacking in certain areas relative to suppliers without identifying and understanding the underlying causes. Such an approach may be appropriate in the case of peripheral activities that add little value in the eyes of customers and for which there is a range of potential suppliers. However, in the case of more important activities a fuller understanding of how the activity is performed is required. This involves fully understanding the critical determinants of successfully performing the activity regardless of whether it is being sourced internally or outsourced to an external supplier. In fact, diagnosing the causes of poor performance can provide valuable insights into how the activity can be improved internally without having to consider outsourcing as an option. This is the most challenging aspect of capability analysis. Although cost analysis and benchmarking are valuable techniques in identifying disparities in performance, considerable judgement is required in understanding the source of superior performance particularly in the context of critical activities. In effect, the sourcing organisation is attempting to understand how superior performance in the activity is achieved.

Porter's (1985) generic drivers (as shown in Table 7.1) associated with either a cost leadership or differentiation strategy can inform this analysis. Table 7.15 shows some examples of how the generic drivers can be a source of advantage in organisational activities. Porter argues that the most important drivers of competitive advantage in an activity include scale, cumulative learning in the activity, linkages between the activity and others, the ability to share the activity with other business units, capacity utilisation, the location of the activity, the extent of

Table 7.15 *Examples of potential drivers of advantage in organisational activities*

Driver	Example
Economies of scale	Can lead to lower costs in a number of areas including production and marketing
Experience	Experience accumulated over time in an activity can be a major source of a strong cost position
Internal linkages	Internal cross-functional collaboration can facilitate better customer service in the form of greater responsiveness to customer needs
External linkages	Effective supplier management can positively impact upon the price of materials
Location	Proximity to most important customers can lead to the provision of a higher level of service at a lower cost
Institutional factors	The influence of government legislation or level of unionisation can impact upon the cost structure of the organisation

vertical integration in performing the activity, institutional factors such as government legislation affecting how the activity is performed, and the policy choices of the organisation on how to configure the activity. Porter (1991) argues that these drivers are the underlying sources of competitive advantage and make competitive advantage operational. These drivers can influence either a cost or differentiation strategy. For example, the development of a strong brand is often considered to be a source of competitive advantage. A strong brand built up over time can lead to a cost advantage – less need for marketing – and also be a source of differentiation – customers are willing to pay a premium price for the product. The generic drivers can be used as a framework for analysing the source of advantage in an activity. Table 4.1, in Chapter 4, illustrates the relationship between activities, types of advantage, potential sources of advantage and resources. For example, consider a garment manufacturer that has focused on a number of specialist niche markets. In order to develop and sustain such a position the organisation has focused primarily on marketing and product design and outsourced garment manufacture to suppliers. The organisation believes that it has a superior position in product design in relation to its direct competitors. The source of this advantage is based upon a number of sources including *experience* in design and *linkages* in the form of cross-functional collaboration between designers, marketing, purchasing and manufacturing. To create this advantage also requires the deployment of a number of resources including equipment, people, finance and relationships.

So far, the discussion in relation to the source of advantage has focused on the activity level. At a more general level, the source of advantage will be influenced by the actions of management in determining the boundary of the organisation. In fact one of the generic drivers refers to vertical integration – the scope and organisation of activities. One of the key objectives in determining the boundary of the organisation is to focus on activities in which the organisation has a particular strength and which it can perform at a lower cost than competitors. Ultimately, the competitive position of the organisation will be influenced by management decisions on a range of areas including which customers it serves, relations employed with suppliers, the allocation and deployment of resources to certain activities, etc. However, analysis at an activity level is crucial to understanding the sustainability of a superior performance position. In the case of critical activities, such an understanding will have a fundamental influence on the sourcing option chosen due to the implications for the competitive position of the organisation. Understanding the source of the advantage will provide a sound indication of the difficulties with attempting to replicate or surpass superior performance levels in the activity. These issues are considered in more detail in the following chapter in the analysis of the sustainability of superior performance in an activity – possessed either by the internal or external source.

REFERENCES

Alexander, M. and Young, D. (1996). Strategic outsourcing. *Long Range Planning*, **29**, No. 1, 116–19.

American Productivity and Quality Center. (1993). *Training Brochure*. Houston, Tx: American Productivity and Quality Center.

Berliner, C. and Brimson, J. A. (1988). *Cost Management for Today's Advanced Manufacturing: The CAM-I Conceptual Design*. Boston: Harvard Business School Press.

Bruck, F. (1995). Make versus buy: the wrong decisions cost. *The McKinsey Quarterly*, **1**, 28–47.

Carpinetti, L. C. R.and de Melo, A. M. (2002). What to benchmark? A systematic approach and cases. *Benchmarking: An International Journal*, **9**, No. 3, 244–55.

Cooper, R. and Kaplan R. S. (1991). Profit priorities from activity-based costing. *Harvard Business Review*, **69**, No. 3, May–June, 130–5.

Drew, S. A. W. (1997). From knowledge to action: the impact of benchmarking on organisational performance. *Long Range Planning*, **30**, No. 3, 427–41.

Eccles, R. and Nohria, N. (1992). *Beyond the Hype: Rediscovering the Essence of Management*. Boston: Harvard Business School Press.

Ellis, G. (1992). Make-or-buy decisions: a simpler approach. *Management Accounting*, June, 22–3.

Foot, J. (1998). *How to Do Benchmarking: A Practitioner's Guide*. London: Inter Authorities Group.

Gunasekaran, A. (1999). A framework for the design and audit of an activity-based costing system. *Managerial Accounting Journal*, **14**, No. 3, 118–126.

Hamel, G., and Prahalad, C. K. (1994). *Competing for the Future*. Boston: Harvard Business Press, pp. 84–5.

Hiltrop, J. M. and Despres, C. (1994). Benchmarking the performance of human resource management. *Long Range Planning*, **27**, No. 6, 43–57.

Kaplan, R. and Norton, D. (1992). The balanced scorecard-measures that drive performance. *Harvard Business Review*, **70**, No. 1, January–February, 71–9.

Karlof, B. and Ostblom, S. (1995). *Benchmarking: A Signpost to Excellence in Quality and Productivity*. Chichester: John Wiley and Sons.

Kennedy, A. (1996). ABC basics. *Management Accounting*, June, 22–24.

Lebas, M. (1999). Which ABC? Accounting based on causality rather than activity-based costing. *European Management Journal*, **17**, No. 5, 501–11.

Murdoch, A. (1997). Lateral benchmarking or what formula one taught an airline. *Management Today*, November, 84–87.

Porter, M. E. (1991). Towards a dynamic theory of strategy. *Strategic Management Journal*, **12**, 95–117.

Porter, M. E. (1985). *Competitive Advantage: Creating and Sustaining Superior Performance*. New York: Free Press.

Senge, P. M. (1990). *The Fifth Discipline*. New York: Doubleday.

Shank, J. K. and Govindarajan, V. (1993). *Strategic Cost Management*. New York: The Free Press.

Sweeney, M. T. (1994). Benchmarking for strategic manufacturing management. *International Journal of Operations and Production Management*, **14**, No. 9, 4–15.

Walleck, A. S., O'Halloran, J. D., and Leader, C. A., (1991). Benchmarking world-class performance. *The McKinsey Quarterly*, **1**, 3–24.

Wisniewski, M. (2001). Measuring up to the best: a manager's guide to benchmarking. In: *Exploring Public Sector Strategy*, London: Financial Times: Prentice Hall, 84–110.

An analysis of the strategic sourcing options

8.1 Introduction

The preceding analysis considered the dimensions of activity importance and organisational capability in the context of outsourcing. These two dimensions yield a number of strategic sourcing options as shown on the matrix in Figure 8.1. This chapter is concerned with evaluating the implications of each of these potential strategic sourcing options. These options should be considered in the context of the following key determinants.

- *The disparity in performance* – this is concerned with determining the sustainability of superior performance in an activity – either by the internal or external source. Some of the insights gained from the capability analysis will assist at this stage. Having a clear understanding of the source of superior performance can provide a reliable indication of the sustainability of such a position. For example, if the company's capabilities lag considerably behind the capabilities of external suppliers, then it may be difficult to justify a substantial investment of resources in order to match or advance upon the capabilities of these external suppliers.

- *Technology influences* – this is concerned with understanding the influence of technology on the choice of sourcing option. For example, if the environment is characterised by rapid advances in technology then any performance advantages the organisation possesses may be difficult to sustain over a long period of time.

- *External considerations* – factors in the external environment such as the political context, market growth rates, the level of competition and barriers to entry should also be considered. For example, in an industry characterised by high levels of growth and competitive rivalry, competitors can rapidly erode a superior performance advantage in an activity.

- *Behavioural considerations* – with outsourcing being considered as an option, there are likely to be a number of behavioural issues that will affect the freedom of the organisation to outsource. For example, there is the potential for significant workforce resistance to such a move particularly from employees that

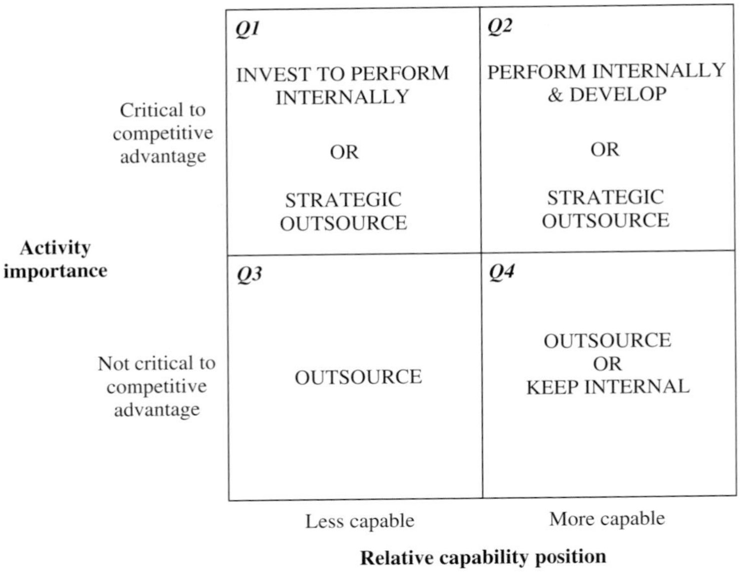

Q1	Q2
INVEST TO PERFORM INTERNALLY OR STRATEGIC OUTSOURCE	PERFORM INTERNALLY & DEVELOP OR STRATEGIC OUTSOURCE
Q3	Q4
OUTSOURCE	OUTSOURCE OR KEEP INTERNAL

Critical to competitive advantage (top rows)
Not critical to competitive advantage (bottom rows)

Activity importance

Less capable   More capable

Relative capability position

Figure 8.1 The Strategic Sourcing Options

are going to be directly affected. In fact, in the case of an organisation that is highly unionised, workforce resistance will act as a powerful inhibitor to any form of outsourcing.

- *Supply market risk* – if outsourcing is being considered, then the level of risk associated with the relevant supply market has to be considered. For example, if analysis of the supply market reveals a high level of rivalry between potential suppliers in the supply market then this is a reliable indicator of low risk and the potential for outsourcing.

These determinants will now be considered in more detail in the following sections along with the conditions that favour the choice of each strategic sourcing option in Figure 8.1.

8.2   An analysis of the disparity in performance

The capability analysis considered the stages involved in analysing the capability of the sourcing organisation in certain activities in relation to external sources. This analysis will have revealed valuable insights into both the significance and the reasons for the disparity in performance. For example, the analysis may have revealed that external suppliers can provide the activity at a much lower cost than internal sources. Moreover, it was important to determine how such superior-performance could be achieved. In the previous example, the superior cost performance of external suppliers may have been as a result of scale economies and greater experience in the process under scrutiny. This type of analysis is crucial to understanding the sustainability of a superior performance position. In the case of critical activities, such an understanding will have a fundamental influence on the sourcing option chosen due to the implications for the competitive position of the organisation. However, whether the organisation is more or less competent than external sources, the same issues have to be addressed when considering whether to perform a critical activity internally. for example, if the analysis has revealed that the internal source is more capable than external sources then the sourcing organisation has to consider how sustainable this performance is likely to be over time. Alternatively, if the analysis has revealed that external sources are more competent, then the sourcing organisation has to determine the degree of difficulty in attempting to replicate and surpass this performance advantage.

Therefore, the central theme of this stage of the analysisis to understand the sustainability of the superior performance in the activity – either by the internal source or external source – over time. It is crucial to understand the source of the advantage because it will provide a sound indication of the difficulties in attempting to replicate or surpass superior performance levels in the activity. Once there is a clear understanding of the source of superior performance in the activity under scrutiny, it is important to determine how readily this superior performance can be replicated. Barney (2002) has identified a number of factors that can provide valuable insights into the potential for replicating superior performance in an activity:

8.2.1   Unique historical conditions

The development of a capability by an organisation may have been as a result of certain unique historical conditions. For example, the ability of an organisation to develop and exploit a capability can depend upon being in the 'right place at the right time'. Over time or under different conditions, organisations that do not possess the capability face a significant cost disadvantage in obtaining and developing such a capability. For example, Caterpillar was able to develop a global service and distribution network for heavy construction equipment as a result of being the primary supplier of this equipment to the Allied forces during the Second World War. After the Second World War this capability in global service and distribution created a significant advantage that was difficult for competitors to replicate. Barney (2002) has identified two ways in which unique historical conditions can make a capability difficult to replicate.

- *First-mover advantage* – this involves an organisation being the first to exploit an opportunity in a market. Being first to market with an innovative business model can increase customer switching costs and develop a strong brand. For example, eBay.com was one of the first companies to exploit the positive feedback dynamics associated with the Internet. eBay.com exploited the positive feedback dynamics of the Internet to develop a capability, which was extremely difficult for competitors to replicate. The company was the first auction site to attract a critical mass of both buyers and sellers. Consequently, with more buyers and sellers connected, the site becomes more valuable and attractive to buyers and sellers.

- *Path dependence* – to develop superior performance in an activity may involve a long and complex learning process. When there is no shortcut or straightforward means of carrying out this process, it is referred to as path dependent (Arthur, 1989). for example, consider a company with a strong quality position in a particular process. Such a capability has been developed over a long period of time through the many interactions of people within the company that are either directly or indirectly responsible for quality of the process. Therefore, it is extremely difficult for a competitor to quickly replicate such a strong position. Furthermore, it is unlikely that such a capability can be procured from the supply market.

8.2.2   Social complexity

A capability may be difficult to imitate because it is composed of socially complex phenomena that other companies find hard to systematically manage and influence. There are a number of capabilities that may exhibit socially complex characteristics. The culture of an organisation is a significant determinant of its ability to respond to changes in the external environment. For example, consider a high technology company with a dynamic culture. Such a culture should facilitate the development of a research and development capability that allows the company to respond more rapidly than competitors to environmental opportunities with innovative new products. Although the value of such a capability is well known to other companies, it can be extremely difficult to create in the short-term. Therefore, when a capability is based on such complex social phenomena, the ability of other companies to imitate such a capability is significantly constrained. The presence of social complexity can also be a barrier to the development of a capability. For example, the presence of a rigid culture within an organisation can reduce the ability of such a company to develop new capabilities in response to changes in the external environment.

8.2.3   Causal ambiguity

Lippman and Rumelt (1982) used causal ambiguity to describe the events surrounding business actions and their consequences that make it difficult for a competitor to replicate superior performance in an activity. Such a capability may be difficult to copy because other companies cannot understand the relationship between the resources and capabilities controlled by the company that possesses the capability. If an advantage is based on capabilities that display causally ambiguous characteristics, then it will be difficult for competitors to imitate such an advantage. Essentially, companies are unable to understand what the capability is and how to create it. Barney (2002) provides a number of reasons for this lack of understanding.

- The resources that create the capability are taken for granted and are just part of the day-to-day running of the company. Itami (1987) defines these types of organisational characteristics as invisible assets. Teamwork amongst management, organisational culture and relationships amongst employees may be 'invisible'.
- The company may have a number of assumptions on how they have developed superior performance in a capability. However, they may not be able to determine which elements either alone or in combination create the capability.
- It may be the case that there are a great number of organisational characteristics that when integrated creates a strong capability in a particular area. In such circumstances it will be extremely difficult to isolate the specific elements that contribute to the development of the capability.

Reed and DeFillippi (1990) have identified three sources that create causal ambiguity for a company both separately and simultaneously.

- *Complexity* – refers to the inter-relationships between skills, and between skills and assets. This is similar to social complexity. The way in which a company integrates its skills and resources can lead to superior performance in certain activities. Complexity arises from the large number of technologies, organisation routines, and both individual and team-based experience. The complexity between the constituent elements that comprise a capability may mean that few individuals can comprehend how superior performance in the activity is achieved. Therefore, causal ambiguity results from this complexity leading to limited potential for competitive replication.
- *Specificity* – refers to the transaction-specific skills and assets that are utilised in the production processes and provision of services for particular customers. Specificity is concerned with the long-term investments that are made to support the particular transactions. Both the buyer and supplier incur transaction-specific investments as well as mutual value in building the relationship. The benefits that are derived from this type of relationship are specific to the buyers and suppliers in the relationship. For example, in a manufacturing context it is possible for a buying organisation to build a relationship with a certain supplier through mutual investments, which leads to the sourcing of lower cost and higher quality components. Competitors of the buying organisation will find it

difficult to leverage the same performance levels from the supplier because of the transaction-specific nature of the relationship between the buying organisation and supplier. Therefore, due to the highly specific nature of the relationship, it is causally ambiguous to competitors.

- *Tacitness* – this is related to the influence of knowledge upon the development of the capability. Specifically, tacitness refers to the implicit and non-codifiable accumulation of skills that are developed through experience and practice. Tacitness is embodied within the skill component of competencies. For example, a strong capability in logistics will involve the integration of skills in a number of areas including planning, scheduling and the application of information technology. The tacit nature of this knowledge makes it extremely difficult for decision rules and protocols that create the capability to be codified and made explicit. Therefore, such tacitness can be a source of causal ambiguity. In fact, the level of ambiguity may be so great that not even the company that possesses the capability comprehends the relationships between its constituent knowledge and skills.

8.2.4 The influence of knowledge

Understanding the influence of knowledge can be important when determining the difficulties of replicating a superior position in an activity. Grant (1996) considers organisational capability as the outcome of knowledge integration. Complex, team-based activities such as American Express's customer billing system and Chrysler's design process are dependent upon the ability of these companies to integrate the knowledge of many individual business specialists. For example, without the application and integration of knowledge, Chrysler's design process is just a collection of resources including staff, equipment and physical infrastructure. Quinn (1992) has argued that the ability to manage knowledge-based intellect is a critical skill. The ability of a firm to sustain a competitive advantage is based upon the knowledge and capabilities of its people (Savage, 1990). The theme of knowledge is often related to the learning organisation. For example, Senge (1990) argues that successful organisations perceive themselves as learning organisations striving to continuously improve their knowledge assets. There are two types of knowledge that can reside within and around an organisation (Nonaka and Takeuchi, 1995).

- *Tacit* – clearly, this is similar to 'tacitness' – knowledge that is difficult to formalise because it normally resides in heads of individuals. Subjective insights, intuitions and hunches are examples of this type of knowledge. Therefore, this type of knowledge is accumulated over time and is a result of the experience of learning by doing. Furthermore, to acquire tacit knowledge may involve sharing and exchanging with partners outside the organisation.
- *Explicit* – this is knowledge that can be represented in words and numbers and readily communicated. For example, explicit knowledge can be easily

documented in procedural manuals and databases. Examples of explicit knowledge include the formal documentation of business processes or recording of internal meetings.

These two types of knowledge are very pertinent to understanding the difficulties associated with replicating a capability. If a capability is based predominantly upon tacit knowledge, then it will be much more difficult for competitors to replicate than one based upon explicit knowledge. The source of strength in this capability may be based upon tacit knowledge that is dispersed across people, processes and locations in a company. For example, the research and design capability of a manufacturing company is based upon the many detailed understandings of the linkages between customer requirements, system requirements and component specifications. Such a capability is unique to the company, and has been developed in numerous interactions between groups of manufacturing, design, and marketing people. The capability is very much based upon the tacit knowledge of people that has been accumulated over time. Moreover, a capability that is predominantly based on tacit knowledge through the complex relationships of people will be more difficult to replicate than a capability that is dominated by physical resources such as machine tools and robots. For example, companies attempting to replicate capabilities based upon physical resources will not have the same difficulties with that of tacit knowledge. Many physical resources can be readily sourced in external supply markets.

As well as considering potential changes in physical resources, the characteristics of the knowledge of people in the organisation that underpins the capability should be considered. The strength of a capability is dependent upon how the organisation accesses and integrates the knowledge of its employees. In an industry where employees are mobile, organisational capability depends primarily upon how the organisation integrates rather than on the breadth of knowledge, which employees possess (Grant, 1996). The higher the level and sophistication of knowledge, whether in the form of language, shared meaning, or mutual recognition of knowledge, the more efficient integration is likely to be. The shift away from expanding specialist knowledge towards increased cross training and job rotation should be based upon the premise that increased common knowledge will enhance organisational capabilities. Nonaka and Takeuchi (1995) argue that successful companies can adapt and expand the knowledge of individuals to create a 'spiral of interaction' between tacit and explicit knowledge via the following four processes.

- *Socialisation* – involves the transfer of tacit knowledge through the sharing of experiences between individuals. For example, socialisation can occur through individuals working in groups or observing others undertaking certain tasks.
- *Externalisation* – is concerned with making tacit knowledge more explicit which in turn makes it more readily accessible throughout the organisation. For example, an

individual may create a model or drawing to create a greater understanding of a particular concept.

- *Combination* – involves combining explicit knowledge from a range of sources into something new. For example, many retailers use intelligent systems technologies to integrate knowledge of customer preferences from a range of sources in order to build up rich customer portraits for target marking purposes.
- *Internalisation* – is concerned with transferring explicit knowledge into tacit knowledge. For example, new staff may be able gain experience by accessing documentation on solutions to common customer problems. This is similar to the process of learning by doing.

The exploitation of knowledge in relation to the needs of customers can also be a source of competitive strength for the organisation. Instead of simply offering a product, many organisations now have started to allocate considerable resource to understanding how their customers use their products or services and provide guidance on how they should be used more effectively. For example, in the airline industry many established airlines have come under intense competition from low-cost airlines such as Ryanair and Easyjet in their short-haul markets. As a result, the established airlines have begun to focus on building relationships with more profitable customers that are categorised as long-haul frequent flyers. Some airlines have begun to exploit their knowledge of customer needs as a potential source of competitive differentiation (McIvor *et al.*, 2003). The established airlines have already been exploiting Internet technologies to capture and integrate critical customer preferences across a range of products and services. Via sophisticated customer profile databases it is now possible to automatically inform staff of customer preferences such as aisle seats or in-flight entertainment. Previously, such information may have been collected but not used to enhance the needs of their most profitable customers. Now, information is being collected from a number of sources and interpreted to create valuable knowledge of customer needs. For example, customer requirements captured at a hotel in one city will affect the services the next time the customer stays with the same hotel chain in another city. Therefore, such information can be captured, analysed, shared, and used to enhance the next customer experience.

8.3   Technology influences

Technology can have a significant influence upon the choice of sourcing option. Changes in technology can allow organisations to develop significant advantages and in some cases change the nature of competition within an industry. Examples of technologies include information technology, robotics, machine tools, etc. Technology of some type is employed in almost all of the activities that an organisation performs. For example, computer-aided design technology is commonplace in most organisations involved in product design and manufacture. In order to examine the influence of technology on outsourcing evaluation, technology should be considered in the following contexts.

- *Deployment of technology* – the deployment of technology by an organisation can have a significant impact upon its competitive position. Innovative new technologies deployed by companies can erode capabilities that have created competitive advantage for their competitors in the past. For example, many high street travel agents have come under intense pressure from online travel agents. Previously, a key source of strength for travel agents in the high street was in the activity of assisting and influencing the customer in the selection and purchase of travel services. The travel agent in the high street derived much of its revenue from performing these functions. However, a plethora of online travel agents such as Expedia.com, Opodo.com and Travelocity.com have employed Internet technologies to perform many of these same functions at a much lower cost.

- *Capability in a technology* – this refers to a technology that an organisation owns and uses in its end products. In outsourcing evaluation, as well as assessing the importance of technology on competitive advantage, companies must assess the superiority of the position of either the internal or external source in the technology under scrutiny. Clearly, the strength of the advantage in the technology will primarily depend upon the difficulties with replicating it. Some of the issues discussed earlier are particularly pertinent to this analysis. The level of causal ambiguity associated with the advantage will affect the ease of replication. For example, if there is a low level of causal ambiguity and high level of competition, it is likely that the advantage will be eroded very quickly. The significance of the disparity in the technology is also a reliable indicator of the strength of the advantage. It is unlikely that a company lagging considerably in a technology can move to a position of leadership. For example, Cummins Engines were faced with a similar situation when they benchmarked their internal capabilities in piston technology relative to their suppliers. They discovered that Cummins' internal design and manufacturing capabilities lagged considerably behind two of those suppliers who possessed world-class capabilities in the technology. These two companies were aggressive innovators. Their scale allowed them to invest more than twenty times as much as Cummins in product and process research and development. Cummins had no choice other than to outsource pistons (Venkatesan, 1992).

The importance of the technology in the future should be considered. For example, an organisation may be a leader in a mature technology that has little impact upon creating value for customers in the future. In this case, it may be more prudent to use a competent supplier in this technology and focus resources on emerging and embryonic technologies that are more likely to be a source of competitive differentiation in the future. Another important influence on outsourcing evaluation is the rate of technological change. As well as being a major source of competitive strength, advances in technology by a competitor can erode a strong position very rapidly. Companies can typically gain detailed knowledge about the new products of their competitors within one year of development (Ghemawat, 1986). The rate of technological change differs across many industries. For example, organisations competing in the semiconductor industry experience much more technological change than organisations that operate in the construction machinery industry. In an industry that is rapidly changing as a result of advances in technology, an organisation must determine whether it can sustain the development and performance in a technology more rapidly than the rate of technological change in the industry.

8.4   External considerations

External factors can have a significant impact upon the outsourcing decision. These factors exist at both the macro and industry level.

8.4.1   Macro-level

Industries and markets are embedded within a wider macro-environment. The macro-environment can include economic, political and legal factors. Changes in these factors can influence the outsourcing decision. Government policy can influence the ease with which companies can outsource. For example, the power of the unions in an economy in many cases is significantly influenced by government legislation. Throughout the 1980s the enactment of legislation in the UK to counter the power of trade unions has made it less difficult to lay off employees than in other European countries. Furthermore, in other European countries like Germany there are higher costs associated with making employees redundant. The presence of more powerful unions can significantly impinge upon the freedom with which companies can outsource. Related to government policy are political changes in the macro-environment such as the collapse of the Soviet Bloc and the liberalisation of international trade. Such changes have created opportunities for large Western companies to outsource production from their domestic base to developing economies in order to access lower labour costs and also gain entry into these markets for their products. These trends have been very pronounced in industries where labour is the most significant cost in the production process. The outsourcing of this type of production has led to the decimation of many high volume labour-intensive industries in many developed economies. Along with labour costs, labour skills can also influence companies in their outsourcing decisions. For example, many companies have begun to outsource much of their information technology requirements to countries such as India and Ireland to access a highly skilled and cost competitive labour market. Advances in information and communication technologies (ICTs) have facilitated this shift of supply to these economies. These countries have also proactively developed their skills base in order to make themselves more attractive to this type of trend.

Illustration 8.1

Developing national competitiveness: the rise of India as an outsourcing destination

In many developed economies outsourcing to offshore locations has often sparked considerable controversy and debate as a result of job losses in the areas affected and concerns on its impact upon international competitiveness. However, this trend has created a significant opportunity for developing economies to develop their national competitiveness and attractiveness for foreign investment from developed economies. One country that has benefited greatly from this trend has been India. Whilst China is regarded a major source for offshore manufacturing from Western companies, India has become the global leader in the service arena. India has become a major outsourcing destination for European and US companies for business services including telemarketing, software development, finance and human resources. For example, in the area of software development India has become the favoured destination for the majority of US companies engaging in offshoring with companies such as General Electric, American Express, AT&T and Citibank outsourcing many of their back office operations to India. India has taken advantage of the trend in both the US and UK of outsourcing service-type activities. Much of this has been as a result of the enormous growth in the service sectors of these economies. In many instances, companies in these countries have had to use offshore locations because of a shortage of skilled labour. Of course, one of the most attractive features of India as an offshore destination has been the lower labour costs relative to Western economies with for example, the labour cost ratio in software development between the US and India at about 8 : 1. India has also benefited from having English as the working language which has given it an advantage over other offshore locations such as China and Russia.

The Indian government has been very proactive in developing its national capabilities in a range of areas in order to enhance its attractiveness to Western companies. In particular, the Indian government has made foreign outsourcing a priority and established development agencies to encourage Western companies to relocate parts of their internal operations and use independent businesses to perform a range of business activities. There has been considerable investment in modern infrastructure including roads, airports, electricity generation, education and telecommunications. The privatisation of the telecommunications sector has significantly reduced telecommunications costs and investments in fibre-optic and satellite technology have facilitated the interfacing of systems and consumers globally. India has also been very proactive in creating awareness amongst Western companies that Indian companies are capable of performing many business activities at comparable or higher levels of performance to their Western counterparts. For example, in the software development sector, the Indian National Association of Software and Service Companies, with support from government, has conducted campaigns to enhance awareness of the sector's capabilities amongst Western companies. In fact, the Indian software sector has built a strong global brand in the software development and services sector through a large English-speaking, highly skilled and lower-cost labour force. Individual firms have also enhanced sectoral competitiveness, which in turn has allowed India to create capabilities that are difficult for other competing off-shore locations to replicate. For example, companies such as InfoSys, one of India's largest technology companies based in Bangalore, has developed tools, methodologies, process and management expertise for providing services to clients across geographic distances. InfoSys segments large-scale software development projects into two categories of tasks. The first part of the project includes tasks that have to be carried out close to the client, whilst the second involves tasks that can be carried out remotely in its technology-enabled development centres in India.

The success of India has also brought its problems. Increasingly, home-based Indian companies have come under competition for skilled labour from Western companies establishing operations in India. Furthermore, domestic providers of services being offshored in developed economies are becoming more competitive in terms of productivity and innovation as they face greater competition from companies in offshore locations such as India. In some cases, Indian companies have failed to achieve the performance levels required by Western companies. For example, in the area of call centre operations, European and the US companies have experienced poor service levels from Indian companies due to cultural and language difficulties. Protectionist practices in some Western economies have also appeared in the form of, for example, many US states enacting legislation to prevent federal government contracts being given to off-shore service providers. However, India is in a strong position to defend its position with some estimating that it is likely to maintain its low labour cost position relative to its competitors for 20 years. Of course, the Indian government is aware that it must build its competitiveness on more than just having a strong relative cost position. Whilst India has focused heavily on the telemarketing and software development areas it is also expanding into other business areas including research and development, engineering and design. This is part of a strategy of 'moving up the value chain' into more value-adding activities in order to move away from being perceived primarily as a low-cost outsourcing destination.

Sources: Kripalani, M. and Engardio, P. (2003). The rise of India. *Business Week*, 8th December, 38–46.
Kumra, G. and Sinha, J. (2003). The next hurdle for Indian IT. *McKinsey Quarterly*, Special Edition: Global Directions, 43–53.

8.4.2 Industry/market level

Industry factors such as the rate of market growth, the intensity of competition and barriers to entry can also impact upon the choice of sourcing option. Each of these factors is inter-related. For example, in an industry characterised by high levels of growth and competitive rivalry, superior performance obtained by a company in an activity can be eroded very quickly. The high level of growth and rivalry forces competitors to innovate in order to equal or surpass the capability. A rapidly growing market may also force a company to outsource certain activities due to internal capacity constraints in order to meet rapid demand for its end products. Factors such as market growth rates, competitive rivalry and entry barriers can also affect the influence of technology on the industry. For example, in a market with a low rate of growth and high entry barriers, a small number of powerful companies may have a 'comfortable' position with little inclination to invest in advances in new technology. Such a market is unattractive to new entrants because of limited potential for profitability. Consequently, this low level of rivalry limits the potential for technological change. Alternatively, markets with high growth rates tend to be characterised by intense rivalry, particularly if barriers to entry are low. The high intensity of rivalry can act as a significant stimulus for companies to design new products or redesign existing products through advances in technology. This in turn accelerates the rate of technological change in the market. Abetti (1989) provides some useful guidelines on the relationship between technology change and characteristics of the market.

- If the rate of technological change is slow, and the rate of market growth is moderate, and there are significant barriers to potential entrants, then the development of the technology internally is the most suitable option. If the development of the technology is successful, the advantage in the technology can be exploited to increase market share.

- If there is a rapid rate of technological change and a low rate of market growth, then there are likely to be risks with developing the technology internally. For example, internal development may lead to the development of an obsolete technology and one for which there is no market. A more suitable strategy in this case is to monitor the various competing technologies and access the most competent source either through acquisition or the development of close collaborative relationship with a supplier.
- If the rate of technological change is slow and the market is growing rapidly, there may not be enough time to develop such a technology internally. In this case, a potential strategy is to licence-in the technology. This should be done on an exclusive basis if the barriers to entry are high, and on a less expensive non-exclusive basis if entry barriers are low.
- If there is a high rate of technological change and market growth is high, a number of options include either acquiring a company or establishing a close relationship with a company that is more competent in the technology.

8.5   Supply market risk

One of the most common fears expressed by managers when considering outsourcing is the risks associated with using an external supplier. When performing an activity internally, companies believe that they can exercise greater control over an internal function and there is less chance of failure of supply than when using an external supplier. For example, companies have natural fears of being held to ransom by a powerful supplier. Furthermore, companies can become overly dependent upon opportunistic suppliers due to a lack of awareness of the issues involved in understanding the supply market. Therefore, when considering outsourcing an activity, it is crucial that a company has an in-depth understanding of the relevant supply market. If supply markets were totally reliable and efficient, companies would outsource all business activities except critical activities that are regarded as a source of competitive advantage. However, most supply markets are imperfect and the buyer must incur some level of risk depending upon the characteristics of the supply market.

Clearly, the level of risk will have a significant influence on the decision to outsource. For example, a high level of risk or uncertainty in the supply market makes keeping the activity internally the preferred option, whereas low a low level risk or uncertainty in the supply market makes outsourcing the preferred option. On the surface, these guidelines appear plausible; however, there are other factors that can influence the decision. For example, consider an organisation that is experiencing high demand for its products in its customer markets. In this case, even though there may be a high level of risk in the supply market, this organisation may have no alternative than to outsource certain activities because it lacks the capacity internally to meet rapidly growing demand for its products. Alternatively, an organisation may develop a new product that involves a number of unique processes that are not accessible in the supply market. Therefore, it has no other choice than to perform the processes internally.

There are a number of reliable indicators of supply market risk.

- *The number of suppliers and buyers* – this is the most important consideration. For example, if the sourcing organisation is considering outsourcing an activity into a supply market where there is only a limited number of competent suppliers then there is the potential for opportunistic behaviour on the part of these suppliers. Such a situation may lead the sourcing organisation to continue to perform the activity internally. Alternatively, a competitive supply market with many competent suppliers provides the sourcing organisation with the opportunity to avail of the associated benefits in the form of lower prices and greater efficiencies. It is also worth emphasising that the number of suppliers that possess the capability to provide the activity is a reliable indicator of the importance of the activity to the sourcing company. For example, if the activity can be readily sourced from a wide range of suppliers then it is not going to be a source of competitive advantage. Competitive demand for supply in the supply market must also be considered, namely the number of buyers. However, along with competitive demand the sourcing organisation must consider its relative size to that of the other buyers. This will affect the level of leverage and service levels that the sourcing organisation can expect to obtain from suppliers. Figure 8.2 illustrates the relationship between the number of buyers and suppliers and its impact upon risk.
- *Uncertainty* – the level of uncertainty in the external environment can create difficulties for an organisation when considering outsourcing. Uncertainty can be in the form of changes in market demand or technology. One of the great difficulties of entering into an outsourcing arrangement is to anticipate whether conditions that held when the agreement was first entered into will still hold in the future. Outsourcing contracts that are drawn up when conditions are uncertain tend to be incomplete. A high level of uncertainty can make it extremely difficult for companies to construct robust contracts that will provide direction as conditions change. Therefore, when the contract has to be renewed or renegotiated, the supplier is likely to behave opportunistically. The supplier may use gaps in the contract to renegotiate terms in its favour. This risk is further compounded if the buying organisation has made significant asset specific investments in the relationship. Such a situation increases the switching costs of moving to another competent supplier in the supply market.
- *Lack of information* – this is related to the availability of information on the general supply market or individual suppliers that again can lead to opportunism.

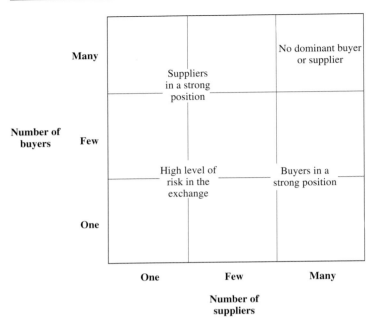

Figure 8.2 The relationship between the number of buyers and suppliers and risk(Adapted from Stuckey and White, 1993)

Lack of information can contribute to opportunism on the part of the supplier in two ways. The first involves *ex ante opportunism* where information is concealed prior to the drawing up of the contract. For example, in outsourcing negotiations a supplier can employ loss-leadership tactics to win business if the sourcing organisation has limited cost information both on the supplier and on the supply market in general. At the most basic level, having access to prices from a range of suppliers in the supply market can lessen the threat of opportunism in the area of loss-leadership by suppliers. However, the sourcing organisation must invest considerable resource in accessing information in order to mitigate against the threats of such opportunism. The second type involves *ex post opportunism* that occurs after the contract has been drawn up and the supplier has concealed information. For example, during the contract the supplier may conceal problems such as labour or raw material shortages that will affect future cost and quality performance.

This section has presented some of the risks associated with sourcing from an external supply market in comparison to an internal supply source. It may be the case that after considering supply market risk, the sourcing organisation feels it is incurring too much risk and decides to continue to perform the activity internally. However, it is worth emphasising that an organisation can incur considerable risks by not outsourcing selected activities. For example, performing too many activities in an organisation's value chain internally can lead to risks in the form of inflexibility and a lack of business focus. It is also possible to reduce some of the supply market risks. For example, developments in information technology now make it possible to have rapid access to information through better communication. In relation to management control, it is often the case that the sourcing organisation has more control over an external supplier than it could ever hope to exercise over an internal function. The sourcing organisation can reduce the risks in the supply market by adopting collaborative relationships with suppliers. For example, long-term contracts, joint ventures, strategic alliances, technology licenses, and franchising are all means of reducing the risks of using external sources.

Illustration 8.2

Virtual integration at Dell

The entry of Dell into the PC industry in the mid-1980s provides an interesting illustration of the impacts of technology, industry characteristics, and supply market risk on the outsourcing decision. In the mid-1980s, the PC industry was dominated by a number of large manufacturers such as IBM, Compaq and Hewlett-Packard. These companies were highly vertically integrated and performed most of the key processes, such as disk drive and memory chip manufacture and application software, internally. Clearly, Dell as a small start-up was not in a position to develop the necessary skills internally in all these technologies, and furthermore, Dell believed that many of these activities created little value for the end customer and could be more readily sourced from the supply market. Instead, Dell pursued a strategy of 'virtual integration' that involved sourcing the relevant technologies from the supply market and developing strong collaborative relationships with their most important suppliers. Virtual integration combines the benefits of two different business models. It offers the advantages of a tightly co-ordinated supply chain that have traditionally come through vertical integration. At the same time, it benefits from the focus and specialisation that drive virtual corporations. Dell's business model has given it the lowest cost base in the industry. Competitors such as Hewlett-Packard and Gateway have struggled to compete because of Dell's ability to reduce costs and enhance value even in times of downturns in customer demand. The level of technological change and industry characteristics has had a significant influence on the reasons why Dell pursued such a strategy.

- In the mid-1980s, the barriers to entry in the PC were lowered which allowed many companies to enter the market and increased the level of competitive rivalry. Therefore, Dell needed to establish a strong market position as

quickly as possible. Many of the components that were used in the manufacture of the PC had become commodities and could be readily sourced
from the supply market. By accessing the capabilities of external suppliers,
Dell was in a position to establish a higher market share than its competitors.
In this context, Dell was able to add capacity more rapidly because it was less
vertically integrated than many of its competitors.

- The life cycles of many of the components in the PC industry are very short –
in some cases a few months. Therefore, it was important to be in a position to
access the most competent external suppliers for a component in terms of
technology, cost and quality. Such suppliers had the scale and experience to
invest more resources than Dell could have done internally. Moreover, the
relationships would last as long as these suppliers maintained their leadership
in the technology involved. Due to the rapid changes in the technologies
involved, if Dell had chosen to develop many of the technologies internally, it
would have incurred the risk of owning potentially obsolete technologies.

Dell's approach to managing its supply chain has allowed it to deliver orders
more rapidly and provide high levels of customer service. Dell sells directly to
customers and bypasses retailers, which allows it to respond more rapidly to
customer orders. The advent of the Internet has also allowed it to further
strengthen this direct-sales approach. One of the key success factors in this
business environment is velocity. Dell prides itself on being able to meet
customer needs within 2 to 3 days of receiving the order. Dell can customise
each computer to the specific requirements of each customer, which it refers to
as the 'build-to-order' model. The success of this model is dependent upon links
with suppliers and the use of information technology. The whole production
process is controlled by software that links its sales and production systems
with a network of suppliers. The company substitutes information for inventory and delivers only when actual demand is received from end-customers. In
some cases, inventory levels and replenishment needs are communicated to
suppliers on an hourly basis. The use of information technology reduces time-
to-market and allows Dell and its suppliers to share information in a range of
areas ranging from simple purchase orders to design databases.

Source: Magretta, J. (1998). The power of virtual integration: an interview with Dell Computer's
Michael Dell. *Harvard Business Review*, **76**, No.2, 73–84.

8.6 Behavioural considerations

Outsourcing involves redrawing the organisational boundaries and changing the
organisational structure. These changes often lead to reductions in the numbers of

employees and changes in the roles and responsibilities of many of the staff remaining in the organisation. As with all major strategies the way in which it is planned and implemented will determine how successful it is. Many outsourcing strategies are based primarily on reducing costs or achieving organisational efficiencies regardless of the damage that can be caused to employees. However, the perspectives and responses of employees at all levels and positions have a significant impact on the successful implementation of outsourcing. Outsourcing often leads to internal fears and resistance from employees. Individual employees, on whom the performance of the organisation is highly dependent, are often largely ignored in planning major change programmes. For example, in the context of downsizing it has been found that many 'survivors' – the employees remaining – no longer trust the organisation and reduce their commitment to the organisation as they believe the whole process to be unfair (Brockner et al., 1994). It has been shown that survivors often display a resistance to change, fear and withdrawal as a result of increased cynicism due to increases in their workloads and unreasonable demands from management (de Vries and Balazas, 1997). Many of these characteristics stem from a lack of employee participation in the process and poor communication that creates a lack of employee trust in top management. For example, survivors will no longer trust top management if they believe information is being withheld and there are inconsistencies between the stated intentions and actions of top management (Mishra et al., 1998). This may also further extend to a lack of confidence in the competency of top management.

Organisations often ignore the fact that the success of strategies that involve change is heavily dependent upon the attitudes and commitment of their workforce (Beaver and Stewart, 1996). Such changes often have a negative impact upon employees. In particular, there are psychological and social implications that can be detrimental to the achievement of the objectives of outsourcing. Also, the trend towards the increased use of 'employee externalisation' arrangements such as part-time, temporary and fixed-term contracts, has led to a change in the relationship between organisations and their employees. As a result, the psychological contract between organisations and the employee in relation to job security in exchange for loyalty has changed (Morrison, 1994). The psychological contract describes the beliefs of the employee in relation to the terms and conditions of the relationship between the organisation and the employee (Rousseau and Greller, 1994). It includes elements such as job satisfaction and promotion prospects in return for a commitment from the organisation. A balanced psychological contract will create an ongoing and close relationship between the organisation and the employee. The balance of the psychological contract is dependent upon the following two conditions (Kakabadse and Kakabadse, 2000a).

- *Reciprocal expectancy* – this relates to the aligning of the expectations of employees of what the organisation will provide with the expectations of the organisation on what it will give to employees (Sims, 1994).

- *Reciprocal exchange* – this depends upon whether or not there is agreement on what is exchanged between the organisation and the employee (Kakabadse *et al.*, 1999).

Clearly, trends such as employee externalisation and outsourcing are impacting upon each of these conditions. In fact, some would argue that many employees no longer believe that a supporting and social contract exists between themselves and their organisation. However, Kakabadse and Kakabadse (2000a) argue that as a result of these developments, a 'new' psychological contract is being forged between the employee and the organisation. For example, the shift from a traditional focus on employment to employability is replacing loyalty with job specialisation. In fact, the trend toward specialisation by organisations is forcing employees to develop expertise and contract with successive organisations as their careers develop. Such developments also emphasise that much of the responsibility for career development rests primarily with the individual rather than with the organisation. In fact, as organisations continue to specialise, outsourcing may present the employee with an opportunity to enhance his or her career. In many cases, employees that are transferred to the supplier have more opportunities to enhance their career. Normally, the employees are transferred to an organisation that specialises in the employee's area of expertise and have better career opportunities than within a small department within the sourcing organisation.

Outsourcing also has major organisation change implications. The changes associated with redesigning job roles and responsibilities are likely to be further emphasised by changing organisation structures. These structures are likely to be flatter, team-based entities whose success is highly dependent upon the establishment of efficient and effective relationships whereby previously regarded 'sensitive' information is shared for mutual benefit with other internal organisational units. These changes not only occur internally but across organisational boundaries. An outsourcing strategy will lead to changes in policies, cultural values, work procedures and processes, relationship between departments, and interactions between buyers and suppliers. In fact, outsourcing can lead to the disintegration of an organisation's culture. The organisation culture is a critical factor that influences successful change (Corby, 1998). Organisation culture is a collection of the values and norms that are shared by the people within the organisation. These values and norms control the way in which people interact with each other internally within the organisation and with external stakeholders. The redrawing of the boundaries of the organisation will impact upon the prevailing values and norms of employees. The culture of the organisation must be considered in the formulation and implementation of the outsourcing strategy.

Often these issues are not addressed when evaluating and implementing the outsourcing strategy. Frequently organisations fail to engage in a process whereby time, money and efforts are invested in bringing about a change in culture, structure and reward systems (Boddy *et al.*, 1998). In effecting major change, difficulties often arise because relatively few organisations establish a link between the objectives set and the resource required. These changes transcend organisational boundaries, and therefore the approach to managing the change process must ensure that complementary activities and behaviours are exhibited within and between organisations. Such change requires a major shift in the mindset and frequently deeply ingrained attitudes and beliefs of organisational members who are likely to have been socialised within an environment that fostered an allegiance to a different value system. Thus, what is required is a multi-functional organisational change strategy that embraces cultural issues and has the support of top management. Baden-Fuller and Stopford (1992) emphasise the need to collaborate across multiple levels in the organisation and highlight the crucial role of middle managers in extending capabilities and developing understanding of priorities. The role of middle managers in this regard is central to widening and deepening the understanding of new ways of operating. Building a climate of change also involves getting people at lower levels "to be able to 'buy in' to the new values in terms that affect the reality of their own work. This also emphasises the need for employee involvement and participation.

Appelbaum *et al.* (1999) have outlined a number of recommendations from a behavioural perspective to managing downsizing that are also pertinent to outsourcing. These are as follows.

- *Human resources as assets versus costs* – many outsourcing strategies are focused primarily on reducing headcount. However, rather than viewing employees as a cost, it is more beneficial to recognise employees as a valuable resource for the organisation. Pursuing this approach involves long-term planning for the development of human resources. Cascio *et al.* (1997) argue that organisations must strategically analyse, restructure and deploy employees on a continuous basis. This involves accounting for all potential costs including those associated with future employment needs.
- *Planning* – as already emphasised, the outsourcing strategy should be part of the overall business strategy of the organisation. The business strategy will include a plan outlining how poorly performing processes can be improved or outsourced whilst focusing on activities that are critical to its competitive position. This plan will also include a plan for retaining vital skills and knowledge (Freeman, 1994). The pursuit of an outsourcing programme has negative consequences in the form of increased levels of stress and a lack of job security. Rather than reacting to these negative consequences, the organisation should have plans in place to alleviate the fears and concerns of employees and create a sense of trust between the organisation and employee. For example, the organisation may be able to re-deploy or retrain employees rather than transfer them to the supplier or make them redundant.

- *Participation* – some effort should be made to involve employees in the evaluation of activities for outsourcing. For example, when analysing the capability of internal activities against suppliers, employees involved in the activity in the sourcing organisation should be given the opportunity to bridge any deficiencies in performance. Through greater participation in the evaluation process, employees are more likely to accept any proposed changes. As well as participation from lower levels in the organisation, top management should also be involved in order to champion and support any strategies put forward.

- *Leadership* – with outsourcing being part of the business strategy of the organisation, top management must display leadership in the form of commitment and participation in the process. The participation of top management indicates to employees that top management are aware of their needs and interests (Appelbaum *et al.*, 1999). This will involve top management being willing to assist employees and answer any questions they may have.

- *Communication* – open and honest communication is another critical element in managing the outsourcing process. It is important to inform employees of their intentions in an accurate and timely fashion. By sharing information with employees and displaying a willingness to communicate, management can create a greater sense of trust and honesty with employees. Mishra *et al.* (1998) argues that such an approach will encourage employees to cooperate and help the company to manage any disruptions that are brought about by the changes. The absence of communication will lead to employees feeling both excluded and demoralised (Smeltzer, 1991).

- *Support for 'victims' and 'survivors'* – support for the victims of outsourcing can be provided in a number of ways including redeployment, training and employee transfers to the supplier. Redundancy should be the last resort in the outsourcing process, although in most cases it is unavoidable. In relation to the survivors, support should be provided in order to prevent them from becoming demoralised. Clearly, with colleagues being transferred to suppliers or being made redundant, surviving employees will have legitimate fears for their future. Mishra *el al.* (1998) have found that surviving employees will act in a way which is consistent with the way the organisation has treated employees that have been made redundant.

These behavioural considerations also have to be placed in the political context of the organisation. So far, the discussion in relation to the choice of strategic sourcing options has assumed that the evaluation and decision making process will be carried out in a rational manner; that is, the decision will be based on achieving the overall strategic goals of the organisation rather than the needs of a functional area or individual. However, in reality political influences within an organisation are likely to have a significant impact upon the evaluation of the outsourcing decision and the management of the outsourcing process. For example, when making decisions that affect the strategic direction of the organisation, managers will often make decisions that will enhance their own personal or functional interests. Organisational politics involves the strategies that individuals employ in order to obtain and use power to influence organisational goals in order to further their own interests and ambitions. Power is extremely important as it indicates the extent to which managers can influence decisions and the overall strategic direction of the organisation. In effect, power is the ability of an individual to persuade or force other stakeholders within and around the organisation to follow a particular course of action. Power can be held formally through for example, an individual's position in the organisational hierarchy. Alternatively, power can be held informally through mechanisms such as control over key organisational resources or control over critical information. The influence of power and politics contrasts significantly with the rational model of management decision-making. For example, under the rational model it is assumed that managers will agree on the strategic objectives of the organisation and the strategies that should be pursued to lead to their achievement.

These considerations are quite pertinent to outsourcing evaluation. The power of the function or indeed the manager of the function under scrutiny for outsourcing evaluation will have a major influence on the ease with which the optimal sourcing option can be chosen. For example, although a function may be significantly under-performing in relation to external sources, the function may exert sufficient power within the overall organisation to prevent outsourcing being chosen as an option. Political motives can also significantly limit the objectivity of the analysis of the capability of the function in relation to external sources, particularly if the analysis is undertaken entirely by the function under scrutiny. Lacity and Hischheim (1995) found in an analysis of information systems outsourcing practices that some functions tended to manipulate the information gathered in analysing their capabilities in order to support their own functional objectives rather than the overall goals of the organisation. The level of power exerted by a function is also likely be determined by its relative importance within the organisation. For example, in the case of a peripheral activity, it is unlikely that the manager or individuals within the function will exert sufficient influence within the overall organisation in order to prevent the optimal sourcing option being chosen. However, in the case of a more critical activity, it is likely that the manager and individuals involved will possess a higher level of power and therefore be in a position to influence the sourcing decision in a way that is compatible with their own interests.

The political dimension pervades all organisations and is clearly an important consideration in the context of outsourcing. Understanding power and politics and the major power brokers within the organisation is an important part of outsourcing evaluation and management. The most appropriate sourcing option may not be chosen because of the presence of powerful coalitions within the organisation who may prevent the optimum choice being made because of their own personal interests. Strong influence exerted at senior management level can create the context for outsourcing evaluation being undertaken so that the overall goals of the organisation are achieved. Indeed, in order to undertake outsourcing evaluation and management effectively, it has already been stressed that it must be aligned with the overall objectives of the strategy of the organisation. Senior management must also possess the necessary power in order to follow through on the most appropriate sourcing option. For example, the chief executive or senior management should possess sufficient power to override the potential of an important organisational function blocking the outsourcing of activities within its remit even though it is deemed to be in the best interests of the organisation. In a study of US companies and their experiences with outsourcing, Useem and Harder (2000) found that leadership was an essential element for effective outsourcing. In particular, they found that an effective organisational context for the exercise of leadership involved both top management support and performance management. Both of these elements enabled the goals of the outsourcing strategy to be achieved. In the case of collaborative relationships being adopted with suppliers, leadership was also found to be an essential requirement. The relative power that the chief executive and senior management possess will dictate the way in which outsourcing evaluation and management is handled. For example, a powerful chief executive can push through an outsourcing strategy with little consultation with lower levels in the organisation. Alternatively, where there are powerful groups within the organisation that are sceptical of outsourcing, a more collaborative and consultative style is required. In fact, the general attitude and posture of senior management within an organisation towards its employees will determine the level of influence that these considerations will have upon outsourcing evaluation and management. An organisation that places considerable efforts on good employee relations is likely to employ a more participative approach to the evaluation and management of the outsourcing process. For example, one of the most damaging effects of outsourcing is the fact that company employees engaged in the activity under scrutiny are being effectively labelled as failures as they are weaker than external suppliers. A potential means of attempting to address this concern in a participative way is to allow employees the opportunity to improve in order to match performance levels of external suppliers.

8.7   The implications of the strategic sourcing options

This section considers the implications of each of these strategic sourcing options shown on the matrix on Figure 8.1. Each of these options is considered in the context of the key determinants already discussed in this chapter. Figure 8.3

QUADRANT ONE

Most appropriate when:

Type of disparity - potential to replicate competitor/supplier performance;
Significance of disparity - low;
Technology - stable technology;
External factors - stable environment;
Supply market risk - high;
Behavioural factors - significant barriers to outsourcing.

Most appropriate when:

Type of disparity - difficult to replicate competitor/supplier performance;
Significance of disparity - high;
Technology - rapidly changing;
External factors - highly competitive environment;
Supply market risk - low;
Behavioural factors - manageable.

QUADRANT TWO

Most appropriate when:

Type of disparity - difficult for competitors to replicate;
Significance of disparity - high;
Technology - relatively stable;
External factors - relatively stable environment;
Supply market risk - high;
Behavioural factors - manageable.

Most appropriate when:

Type of disparity - straightforward for competitors to replicate;
Significance of disparity - low;
Technology - rapidly changing;
External factors - highly competitive environment;
Supply market risk - low;
Behavioural factors - manageable.

QUADRANT THREE

Most appropriate when:

Type of disparity - difficult to replicate;
Significance of disparity - very significant;
Supply market risk - low;
Behavioural factors - manageable.

N.B. The key conditions that will prevent outsourcing in this case will be a high level of internal resistance and/or high supply market risk.

QUADRANT FOUR

Most appropriate when:

Type of disparity - straightforward for suppliers to perform;
Significance of disparity - low;
Technology - rapidly changing;
External factors - highly competitive environment;
Supply market risk - low and possible to develop the capabilities of suppliers to undertake the activity;
Behavioural factors - manageable.

Most appropriate when:

Type of disparity - difficult for suppliers to perform;
Significance of disparity - very significant;
Supply market risk - no capable suppliers available and no potential for supplier development;
Behavioural factors - manageable.

N.B. The key conditions that will influence this sourcing option will be the lack of available suppliers and internal resistance to outsourcing.

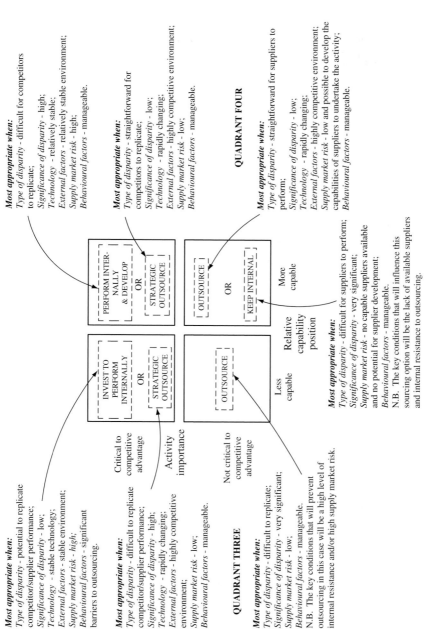

	PERFORM INTER-NALLY & DEVELOP
INVEST TO PERFORM INTERNALLY	OR
OR	STRATEGIC OUTSOURCE
STRATEGIC OUTSOURCE	

Critical to competitive advantage

Activity importance

Not critical to competitive advantage

OUTSOURCE	OUTSOURCE
	OR
	KEEP INTERNAL

Less capable   Relative capability   More
                     position   capable

Figure 8.3 Suggested ideal conditions for the choice of strategic sourcing options

outlines the conditions that are most appropriate for the selection of each of the sourcing options.

8.7.1   Quadrant one

In this quadrant, there are external sources that are more capable than internal ones within the sourcing organisation for a critical activity. The key questions that must be considered in this quadrant are the following.

- How significant is the disparity in performance between the sourcing organisation and potential external sources?
- How much investment is required by the sourcing organisation to equal and surpass the capabilities of external sources?
- What are the behavioural influences on attempting to bridge the disparity in performance?
- Is it possible to outsource the activity and leverage the capabilities of external suppliers?
- Are there internal and external constraints on pursuing such an outsourcing strategy?
- Will customers of the sourcing organisation recognise a difference in its end products if the activity concerned is outsourced?

These questions are examined in the context of the following options associated with this quadrant.

Invest to perform internally
This option involves investing the necessary resources to bridge the disparity between the sourcing organisation and the more competent external providers of the activity. The selection of this option will depend upon the following.

- *Significance of the disparity* – if the disparity is not significant then there is the potential to invest resources in order to perform the activity internally. For example,this option may be desirable in a case where the technologies involved in the activity are in the early stage of development and therefore may offer considerable scope for future growth. However, if the company's capabilities lag considerably behind the capabilities of external providers, then it may be difficult to justify a substantial investment of resources in order to match or advance upon external capabilities.
- *Type of disparity* – the type of disparity is crucial in determining whether it is feasible to invest the necessary resources to match the superior performance of external sources. For example, if the superior cost performance of a supplier is based upon economies of scale then it will be very difficult for the sourcing organisation to achieve such an advantage. Alternatively, analysis of the

activity may reveal that the disparity in performance is in an area such as quality or productivity, which can be addressed through an improvement initiative. An effective benchmarking exercise should also assist in determining what actions need to be taken in order to bring the performance up to a comparable level with external sources.

The *Invest to perform internally* option is most appropriate when the sourcing organisation is in a strong position to bridge the disparity in performance. It is important to emphasise that the organisation may have no other choice because there is considerable risk in using an external supplier for such a critical activity. Furthermore, the presence of internal constraints such as work force resistance and the threat of industrial action may force the sourcing organisation to attempt to improve performance through an internal improvement initiative.

Strategic outsource

This option is likely if the organisation has decided that it is not possible to attempt to bridge the disparity in performance. For example, consider an automaker that has previously designed and manufactured the engines for all its models internally. The same automaker considered engine design and manufacture to be a critical activity. However, through an extensive benchmarking exercise it determined that it no longer possessed the design and manufacturing skills and resources necessary to match the performance levels achieved by other car manufacturers. Even though perceived as a critical activity, it had no other choice than to source engines for a number of its models from another automaker.

The strategic outsource option can be employed to redirect resources to other innovative areas within the organisation. For example, in the 1970s Hotpoint outsourced some of its production needs to Zanussi because its own plants were outdated and had poor productivity levels (Baden-Fuller et al., 2000). Hotpoint then redirected its engineers to develop new production techniques in potentially more innovative areas including the use of pre-painted steel. In particular, developing a capability in pre-painted steel allowed Hotpoint to achieve flexibility in terms of scale and product variety, as well as having the lowest cost base in the domestic appliance industry.

The importance of the activity in the future should also be considered. For example, the organisation may consider outsourcing such an activity, which is likely to diminish in importance in order to focus resource and effort on activities that promise to be a source of competitive differentiation in the future. In certain circumstances the organisation may have no choice other than outsourcing because of internal capacity constraints. Strategic outsourcing is most appropriate when the organisation feels the advantage the external source has in the activity is too difficult to replicate. Conditions in the external environment may favour such an approach. For example, if the organisation is operating in an environment that is experiencing considerable change – either through advances in technology or increased competition – it may be more prudent to outsource certain activities rather than incur the risk of owning too many activities that may hinder future growth.

Acquisition

Another potential option is to acquire a company that possesses superior performance in the activity. For example, the sourcing organisation may have decided that it cannot achieve the required performance levels internally and there is a high level of risk associated with using an external supplier. Pursuing an acquisition strategy has a number of advantages. It is possible to achieve operational synergies through acquisition (Haberberg and Rieple, 2001). This may involve the sharing of distribution channels, the maximisation of manufacturing plant utilisation, the shared use of systems, and infrastructure. Also, acquisition can serve as a rapid, low-cost route into a new product area through the accessing of existing knowledge on both technology and the market. However, the merits of the acquisition option must be weighed against the limitations. Acquisition is an option open only to large organisations. For example, acquisition can be costly in terms of the financial costs involved in making the acquisition. Again, the costs of ownership must be considered – the risks associated with managing and developing too many activities internally. Moreover, acquisition may not be possible because of legal constraints and local ownership rules (Barney, 2002).

8.7.2   Quadrant two

In this quadrant, the organisation is more competent than any other potential external sources in a critical activity. The key questions that must be considered in this quadrant are the following.

- Should the organisation strive to maintain and build upon its superior capability in the activity concerned?
- What are the behavioural influences on maintaining and building the activity internally?
- How significant is the disparity in performance between the sourcing organisation and potential external suppliers?
- Does the organisation possess the necessary resources to provide the activity on an on-going basis?
- Is it possible to outsource the activity and leverage the capabilities of external suppliers?

- Are there internal and external constraints on pursuing such an outsourcing strategy?
- Will customers of the sourcing organisation recognise a difference in its end products if the activity concerned is outsourced?

With this situation, the organisation has the following two possible options with this quadrant.

Perform internally and develop

This strategy involves continuing to perform the activity internally. Again, as with quadrant one it is important to consider both the significance and type of disparity in performance in the activity. For example, if the sourcing organisation has built up a significant performance advantage through scale economies over time, then it is going to be difficult for external sources to replicate such a capability. It is also important to assess whether the organisation can build upon the current advantage by further developing the capability in the activity in order to minimise the risk of external sources matching internal performance levels. For example, a potential constraint to internal development is a lack of skilled labour or financial resources. Clearly, keeping the activity internal is the most appropriate when the sourcing organisation is in a strong position to sustain its performance advantage over time. Moreover, it may not possible to outsource such an activity because of lack of suppliers in the supply market that can meet the performance levels required in the activity.

Strategic outsource

Ideally, an organisation wants to have superior performance in as many of its critical activities as possible. However, it is only possible to possess leadership in a limited number of activities due to the resources required to maintain such a position. In some circumstances, it is possible to gain competitive advantage by outsourcing the assembly or integration of the component parts that comprise a critical activity. Competitive advantage can be achieved in the activity of specifying and integrating external services and other purchases, rather than in assembly and production of the goods themselves. Volkswagen have pursued a similar strategy in their truck and bus plant in Brazil (Simonian, 1996). It has created a new style of relationship with its suppliers. Instead of buying parts from outside, Volkswagen has asked component makers to fit their products directly on the production line. Suppliers no longer simply make parts, but design and develop entire sub-assemblies. Suppliers have assumed responsibility for putting together and installing four 'modules' in all: the chassis; axles and suspension; engines and transmissions; and driver's controls. These 'modules' are regarded as critical components. However, Volkswagen believes that it can benefit more by developing collaborative relationships with its suppliers. It can share costs such as research and development spending for component modules with its suppliers. Volkswagen is acting, in effect, as a 'systems integrator', co-ordinating its suppliers and taking exclusive responsibility only for marketing and selling vehicles. However, it is unlikely that Volkswagen would have the freedom to pursue such a strategy in an established economy like Europe where all the stakeholders, such as unions and suppliers have established positions. Nike has pursued a similar strategy to some extent in its efforts to create maximum value by concentrating on pre-production (research and development) and post-production activities (marketing, distribution, and sales).

Illustration 8.3

Modular production in the automotive industry

The concept of modularisation has become very prominent in many industries. For example, modularisation is very prevalent and necessary in both the engineering and software fields. In particular, many automotive manufacturers such as Volkswagen and Jaguar have been adopting modular-type arrangements with their suppliers. These manufacturers have been attempting to streamline their supply chains and access the capabilities of suppliers. As a result, they have been pursuing the following strategies:

- rationalising their supply bases;
- defining a new set of supply requirements including global sourcing, full service supply and design for assemblies; and
- outsourcing activities traditionally carried out in-house.

The adoption of modular-type supply arrangements has been an integral part of this strategy. Modules refer to a physical assembly such as a seat, dashboard, windshield wiper system or gearbox. For example, a windshield wiper system is composed of a set of electrical components (including resistors, capacitors, transistors, semiconductor chips, and a PCB), mechanical components (including housing and connectors) and chemical components (including silicon, epoxy and solder). In a modular arrangement the module supplier is responsible for assembling the module directly on the automotive manufacturer's assembly line with the manufacturer and supplier working side-by-side. This arrangement is characterised by a long-term contractual arrangement between the manufacturer and first-tier supplier. As part of this arrangement the supplier assumes responsibility for the assembly of the module, installation of module into the vehicle on the assembly line, an investment stake in the operation and management of the suppliers who provide components for the module. The automotive manufacturer provides the plant and assembly line and assumes responsibility for plant co-ordination and final testing. In the case of the manufacture of the Smart

Mincar, this has led to the establishment of a green-field site with the final assembler and all the module suppliers being present on the same location.

A key motive for the move towards modular supply has been to achieve cost reductions. For example, suppliers often have lower labour rates, lower overheads, greater economies of scale whilst the manufacturer reduces inventory and simplifies the transaction with the supplier. The automotive manufacturers also believe they can leverage the capabilities and knowledge of these suppliers in order achieve both performance and design improvements. Through working closely with a limited number of suppliers, the mutual sharing of objectives and resources can be facilitated which leads to more rapid design and manufacture. For example, Skoda has pursued a modular arrangement with a number of suppliers in its Octavia plant. In particular, its relationship with Johnson Controls, the seating manufacturer, has allowed both companies to blend their distinctive competencies in seat design in order to develop a seat which contributes to the overall comfort and well-being and safety of the driver and passenger.

The adoption of modular production arrangements represents a major departure from the highly vertically integrated arrangements traditionally pursued by many Western car makers. By pursuing modular arrangements these organisations have been attempting to combine some of the benefits of vertical integration without financial ownership. By locating suppliers nearby and sharing investments they are reducing the risks associated with being vertical integration whilst having greater control through close communication and high levels of information exchange. The automotive manufacturer can benefit through shared investments in new plant and new products. In turn, suppliers will be able to obtain higher levels of business and become a strategic resource for the automotive manufacturer. However, managing the relationship with modular suppliers presents automotive manufacturers with a challenge in the form of how best to leverage and blend the capabilities of each party into a mutually rewarding relationship. The rationalisation of the supply chain also has the potential to alter the balance of power in the relationship between the automakers and their suppliers. For example, as a result of mergers and acquisitions in areas such as airbags, braking systems, and seat manufacture, supply is concentrated in the hands of a limited number of global companies such as Bosch and Johnson Controls. Indeed, this potential threat could become even more severe if these specialist suppliers engaged in mergers or acquisitions that allowed them to offer a range of modules to automakers.

Sources: Collins, R., Bechler, K. and Pires, S. (1997). Outsourcing in the automotive industry: from JIT to modular consortia. *European Management Journal*, **15**, No.5, 498–508.
Hsuan, J. (1999). The impacts of buyer–supplier relationships on modularisation in new product development. *European Journal of Purchasing and Supply Management*, **5**, 197–209.

8.7.3   Quadrant Three

In this quadrant, there are external suppliers that are more capable than internally within the sourcing organisation for an activity not critical to business success. These activities are suitable candidates for outsourcing. This quadrant can include many straightforward activities, products, services required by the sourcing organisation. Many companies fail to appreciate the opportunity costs of investing in activities that are not critical to business success. For example, in a manufacturing context continuing to produce a component internally may require considerable management and engineering resource. In fact, due to cultural and historical issues, there may be a prevailing view that everything can be done in-house. Attempting to perform too many activities in-house inevitably leads to a situation where no activity is given the sufficient level of attention to create superior performance. Such an attitude may also be reinforced by a culture that has failed to appreciate the capabilities of external suppliers that can perform the same activities more competently. The key questions that must be considered in this quadrant are the following.

- Is there a sufficient number of suppliers in the supply market to ensure adequate competition for the sourcing of the activity?
- To what extent do potential external suppliers of the activity possess proprietary technology that gives them an advantage over other suppliers?
- Is there the potential to outsource the activity and leverage the capabilities of external suppliers?
- Are there internal and external constraints on pursuing such an outsourcing strategy?
- Will customers of the sourcing organisation recognise a difference in its end product if the activity concerned is outsourced?

In this quadrant, the most significant influences are the level of supply market risk and the constraints that impact upon the freedom of the organisation to outsource. For example, if there are only a limited number of capable suppliers in the supply market, the sourcing organisation may decide to continue to perform the activity internally and furthermore, internal constraints such as the threat of industrial may impinge upon the freedom to outsource.

8.7.4   Quadrant Four

In this quadrant, the organisation is more competent than potential external sources in an activity that is not critical to business success. The key questions that must be considered in this quadrant are the following.

- Is there the potential to outsource the activity and leverage the capabilities of external suppliers?

- Are there internal and external constraints on pursuing such an outsourcing strategy?
- Will customers of the sourcing organisation recognise a difference in its end products if the activity concerned is outsourced?

Although the sourcing organisation is more competent than external sources, the activity is not central to competitive advantage. Therefore, the organisation should consider outsourcing such an activity and focusing resources on building capabilities in activities that are more critical to the success of the organisation. However, if the organisation decides to outsource then it will have to develop the capabilities of a supplier to the level achieved internally. There are a number ways of achieving this including a supplier development programme, through the transfer of employees and equipment to a suitable supplier or a management buy-out of the activity under consideration. However, the organisation may decide to continue performing the activity internally. For example, as with many outsourcing decisions there may be internal constraints such as workforce resistance and the costs of moving the activity to an external source. Alternatively, the organisation may decide to divest itself of these types of activities in order to focus on other activities that show greater potential for future growth. For example, Vickers, the UK engineering-based group, sold off Rolls Royce Cars and Cosworth in 1998 in order to generate funds for acquisitions. Vickers re-focused their business in order to become a high-quality marine engineering group. This decision was based on the belief that sea transportation in Europe will become more important as a result of growing congestion on road and rail networks (Kakabadse and Kakabadse, 2000b).

REFERENCES

Abetti, P. A. (1989). *Linking Technology and Business Strategy*. New york: The President's Association, American Management Association.

Appelbaum, S. H., Everard, A. and Hung, L. T. S. (1999). Strategic downsizing: critical success factors. Management Decision, **37**, No. 7, 535–52.

Arthur, W. B. (1989). Competing technologies, increasing returns, and lock-in by historical events. Economic Journal, **99**, 116–31.

Baden-Fuller, C. and Stopford, J. M. (1992). *Rejuvenating the Mature Business*. London: Routledge.

Baden-Fuller, C., Targett, D. and Hunt, B. (2000). Outsourcing to outmanoeuvre: outsourcing re-defines competitive strategy and structure. *European Management Journal*, **18**, No.3, 285–95.

Barney, J. A. (2002). *Gaining and Sustaining Competitive Advantage*. New Jersey: Prentice Hall, second edition.

Beaver, G. and Stewart, J. (1996). HRM: the essential ingredient for successful strategic change. Editorial Journal of Strategic Change, **5**, No. 6, 307–8.

Boddy, D., Cahill, C., Charles, M., Fraser-Jraus, H. and MacBeth, D. (1998). Success and failure in implementing supply chain partnering: an empirical study. *European Journal of Purchasing and Supply Management*, **2**, No. 2/3, 143–51.

Brockner, J., Konovsky, M., Cooper-Schneider, R, Folger, R., Martin, C. and Bies, R. (1994). "Interactive effects of procedural justice and outcome negativity on victims and survivors of job loss. *Academy of Management Journal*, **37**, No. 4, 397–409.

Cascio, W. F., Young, C. E., and Morris, J. R. (1997). Financial consequences of employment-change decisions in major US corporations. *Academy of Management Journal*, **40**, No. 5, 1175–89.

Corby, S. (1998). Industrial relations in civil service agencies: transition or transformation? *Industrial Relations Journal*, **29**, No. 3, 194–207.

De Vries, M. and Balazas, K. (1997). The downside of downsizing. *Human Relations*, **50**, January, 11–50.

Freeman, S. J. (1994). Organisational downsizing as convergence or reorientation: implications for human resource management, *Human Resource Management*, **33**, 2, 213–38.

Ghemawat, P. (1986). Sustainable advantage, *Harvard Business Review*, **64**, No. 5, 53–8.

Grant, R. M. (1996). Toward a knowledge-based theory of the firm. *Strategic Management Journal*, **17**, Winter Special Issue, 109–22.

Haberberg, A. and Rieple, A. (2001). *The Strategric Management of Organisations*. London: Financial Times–Prentice Hall.

Itami, H. (1987). *Mobilising Invisible Assets*. Cambridge, MA: Harvard University Press.

Kakabadse, A, Kakabadse, N. and Kouzmin, A. (1999). The changing nature of the psycho-logical contract and the consequences. In: *Public Service Management: Achieving Quality Performance in the 21ˢᵗCentury*, Ahmed. M. (ed.).Manila: *Eastern Regional Organisation for Public Administration(EROPA) and Public Services Department*, pp. 450–72.

Kakabadse, N. and Kakabadse, A.(2000a). Critical review – outsourcing: a paradigm shift. *The Journal of Management Development*, **19**, No. 8, 670–728.

Kakabadse, N. and Kakabadse, A. (2000b). Sourcing: new face to economies of scale and the emergence of new organisational forms. Knowledge and Process Management, **7**, No. 2, 107–18.

Lacity, M. C. and Hirschheim, R. (1995). The role of benchmarking services in insourcing decisions. In: *Beyond the Information Systems Outsourcing Bandwagon: The Insourcing Response*. Chichester: Wiley.

Lippman, S., and Rumelt, R. P. (1982). Uncertain imitability: an analysis of inter-firm differ-ences in efficiency under competition. *Bell Journal of Economics*, **13**, 418–38.

Magretta, J. (1998). The power of virtual integration: an interview with Dell Computer's Michael Dell, *Harvard Business Review*, **76**, No. 2, 73–84.

McIvor, R. T., O'Reilly, M. D, Ponsonby, S. (2003). The impact of Internet technologies on the airline industry: current strategies and future developments. *The Journal of Strategic Change*, **12**, No.1, 31–47.

Mishra, K. E., Spreitzer, G. M. and Mishra, A. K. (1998). Preserving employee morale during downsizing. *Sloan Management Review*, **39**, No. 2, 83–95.

Morrison, D. (1994). Psychological contracts and change. *Human Resource Management*, **33**, No. 3, 72–84.

Nonaka, I. and Takeuchi, H. (1995). *The Knowledge-Creating Company*. Oxford: Oxford University Press.

Quinn, J. B. (1992). *Intelligent Enterprise: A Knowledge and Service-Based Paradigm for Industry*. New York: Free Press.

Reed, R. and DeFillippi, R. J. (1990). Caususal ambiguity, barries to imitation,and sustainable competitive advantage. *Academy of Management Review*, **15**, No. 1, 88–102.

Rousseau, D. and Greller, M. (1994). Human resource practices: administrative contract makers. *Human Resource Management*, **33**, No. 3, 72–81.

Savage, C. M. (1990). *Fifth Generation Management: Co-creating through Virtual Enterprising, Dynamic Teaching, and Knowledge Networking*. Newton: Butterworth-Heinemann.

Senge, P. M. (1990). *The Fifth Discipline*. New York: Doubleday Currency.

Simionian, H. (1996). Prophet of the production line. Financial Times, London, 29 March, p. 19.

Sims, R. (1994). Human resource management's role in the clarifying the new psychological contract. *Human Resource Management*, **33**, No. 3, 37–42.

Smeltzer, L. R. (1991). An analysis of strategies for announcing organisation-wide changes. *Group and Organisation Studies*, **16**, No. 1, 5–24.

Stuckey, J. and White, D. (1993). When and when not to vertically integrate, *Sloan Management Review*. **34**, No. 3, 71–83.

Useem, M. and Harder, J. (2000). Leading laterally in company outsourcing. *Sloan Management Review*, **41**, No. 2, 25–36.

Venkatesan, R. (1992). Strategic sourcing: to make or not to make. *Harvard Business Review*, **70**, No. 6, 98–107.

One of the major criticisms of outsourcing is that it is an operational tool and cannot be used as a means of building competitive advantage for an organisation. Proponents of this view argue that any activities sourced from suppliers in the supply market are also available to the competitors of the sourcing organisation. In the past, organisations have only outsourced those activities that were regarded as of peripheral concern to an organisation such as cleaning, catering and security and therefore had little impact upon the competitive position of the organisation. However, organisations are now outsourcing more critical activities such as design, manufacture, marketing, distribution and information systems. As a result of outsourcing more critical activities, organisations have been attempting to develop collaborative relationships with suppliers as they seek to reduce the risks associated with outsourcing. By employing astute relationship management mechanisms it is possible for organisations to leverage the capabilities of suppliers in a more effective way than competing organisations. Organisations are now beginning to recognise the benefits of adopting approaches for managing across and beyond the boundaries of the organisation. There is a growing awareness that competitive advantage can be based as much upon managing beyond the ownership boundaries of the organisation rather than on management within those boundaries. Competitive advantage can be achieved through the cultivation and nurturing of linkages between the sourcing organisation and its key suppliers. These benefits are now achievable with the availability of specialists in particular activities that can create a greater depth of knowledge and be much more efficient than integrated companies. For example, there are now specialist service providers in many areas including logistics, human resource management, customer service and support and information systems. Clearly, these trends place supply management at the heart of the outsourcing process. Therefore, it is essential to recognise that activities can be strengthened and developed across organisational boundaries.

This chapter outlines the steps involved in building relationships across organisational boundaries to achieve the outsourcing objectives. The previous stages in the analysis identified the suitability of certain activities for outsourcing in relation to their importance and the capability of the sourcing organisation in relation to external sources – either suppliers or competitors. This analysis led to the development of a number of potential sourcing options including invest to keep the activity in-house; keep the activity in-house and develop; or outsource the activity. These sourcing options were influenced by a number of factors including the nature of the disparity in performance, technology, external and internal influences and supply market risk. It was argued that, where possible non-critical activities should be outsourced. It was also argued that in certain circumstances it is appropriate to outsource critical activities. This chapter outlines the actions that should be taken in the outsourcing process for each of these potential scenarios. Choice of relationship type and management of the relationship is crucial to the development of an effective outsourcing strategy. The choice of the relationship type must reflect the overall objectives for outsourcing the activity. For example, if the key objective of outsourcing is to integrate the capability of a supplier in a critical process into the end-products of the buyer, then a collaborative relationship is likely to be the most appropriate choice. Some of the issues already considered can assist at this stage. For example, the level of supply market risk will have an influence on the type of relationship the organisation will adopt with the supplier. Once the choice of relationship type is made the way in which the buyer – supplier relationship is managed must be based upon the attainment of the objectives set for outsourcing the activity.

Illustration 9.1

Insights on global outsourcing from the toy industry

Eric Johnson carried out a study of supply chain management practices in the toy industry. In the toy industry, supply chains now extend globally and also include many developing countries with the associated currency and political risks that can lead to cost changes and disruptions to supply. Some interesting insights are provided into how toy companies such as Hasbro and Mattel employ outsourcing and supplier relationship management to manage the supply-side risks. Traditionally, toy manufacturers have developed considerable skills in managing the needs of consumer demand. For example, many toy products are comparable to fashion-type products with very short lifecycles, significant markdowns, and rapid upsurges and downturns in demand. However, supply chain management has also become a major challenge as they have rapidly moved production from the US and Europe to countries with lower labour rates such as China, Malaysia and Thailand. In the toy industry, oversupply not only leads to markdowns and write-offs but can also destroy demand for fad-sensitive products. However, scarcity of a product can be a powerful selling proposition during times of peak demand such as Christmas. As well as being challenges there are considerable risks associated with managing both product demand (such as seasonality, volatility of fad-type products, and short product lifecycles) and product supply (such as manufacturing, capacity, currency and political risks associated with employing foreign suppliers).

After the Second World War, many of the toy companies moved their own manufacturing operations to developing economies in Asia such as Malaysia, Thailand and China. However, by moving their own manufacturing operations to these countries, they lost supply chain flexibility due to the length of the distribution channels, communication problems, trade barriers, and pre-ordering products months before they are required. Due to these problems and the risks associated with establishing and managing their own manufacturing plants, both large and small toy makers used outsourcing to meet much of their product requirements. The smaller toy companies outsource all of their manufacturing requirements, whilst the larger toy companies use a combination of wholly owned and outsourced manufacturing sources. Indeed, outsourcing has become a competitive necessity for many of the small toy companies as it allows them to establish a market presence without possessing any manufacturing experience or capital investment. The toy companies believe limited value can be generated in the manufacture of many of these products. The toy companies can focus on generating new ideas and developing new products whilst allowing the suppliers to manufacture the products. As well as outsourcing manufacturing, the toy companies have outsourced logistics services. The manufacturers produce products ready to retail that are sent directly to the retailers with limited intervention from the toy companies.

In relation to supplier management practices in Asia, many of the toy companies use hubs or intermediaries to manage the relationship between the manufacturers. Most of these hubs are financed and managed through Hong Kong joint ventures. For example, Mattel established a new division called Vendor Operations Hong Kong (VO). This division manages a network of around thirty suppliers. A small number of these suppliers are regarded as strategic partners as they account for a significant percentage of their total volume requirements. This division allows Mattel to source a wide range of toys with short lifecycles whilst avoiding the capital requirements associated with internal manufacturing. It also uses these suppliers as additional capacity for its own manufacturing plants particularly in the case of products that have volatile demand. In this case, the toy companies are shifting much of the risks associated with demand uncertainty on to the suppliers. Retailers are constantly demanding shorter lead times from the toy makers. Through outsourcing products with high demand uncertainty, toy makers are in a position to access the capabilities of suppliers and introduce more new products to market. The supplier can reduce these risks by manufacturing a range of products for different toy companies. For example, a downturn in demand for one product may be compensated by an upsurge in demand for another. The toy companies also spread their demand requirements around their supply base in order to create competition between suppliers and allow supply to be switched from poor performing suppliers.

Source: Johnson, M. E. Learning from toys: lessons in managing supply chain risk from toy industry. *California Management Review*, 43, No. 3, 106–124.

9.2   Establish objectives for outsourcing

Many companies embark upon outsourcing without establishing clear objectives on what they intend to achieve. The absence of clear objectives can create difficulties in managing the outsourcing process in a number of areas including selecting the most appropriate supply relationship, drawing up the contract and managing the relationship with the supplier. Establishing clear objectives at the outset will assist in the management of a more effective outsourcing strategy. Typical objectives for outsourcing can include reducing the costs of sourcing the activity, enhancing quality levels and obtaining higher levels of service in the provision of activity. The objectives set will reflect the underlying motives for the organisation considering outsourcing as an appropriate strategy. For example, in many cases, organisations outsource with the objective of obtaining higher quality service at a lower cost from external suppliers. In this case, the key objectives will provide valuable direction for managing the outsourcing process. This will involve the organisation ensuring that it is selecting the most capable supplier in terms of cost and quality. The organisation must also ensure that the chosen supplier maintains and improves its performance throughout the life of the relationship. The objectives for outsourcing are important from a number of perspectives.

- *Selecting the supply relationship* – the objectives established for outsourcing will assist in selecting the type of supply relationship that should be adopted. For example, if the organisation is focusing primarily on attaining the lowest price, then an adversarial relationship is likely to be the most appropriate. Alternatively, if the organisation wishes to access and integrate the knowledge and design skills of the supplier into its own products then a more collaborative relationship is required.
- *Monitoring supplier performance* – once the relationship is established the performance of the supplier must be assessed on an ongoing basis. The

outsourcing objectives will inform the metrics used to assess supplier performance throughout the life of the relationship. For example, if enhanced service
is the key objective for outsourcing, then clearly there must be a number of
metrics used to measure performance in this area.

- *Monitoring the nature of the supply relationship* – the objectives of outsourcing
  influence how the relationship with the supplier is managed. It is important to
  monitor whether the relationship is meeting the overall outsourcing objectives.
  Conditions in the supply market can change which may affect the relationship
  with the supplier. For example, the entry of more competent suppliers into the
  supply market for a particular activity can lead to the buyer having to switch to
  another source of supply in order to continue to meet the outsourcing objectives
  established at the outset.

- *The competitive position of the sourcing organisation* – it has already been argued
  that it is possible for organisations to use outsourcing to contribute to competitive advantage through the development of relationships with suppliers. It is
  important that the outsourcing objectives are linked with the strategic objectives
  of the organisation. This is particularly important in the case of critical activities. In relation to critical activities, some of the objectives for outsourcing must
  reflect the overall strategic goals of the organisation. Although, objectives that
  can be achieved in the short-term for outsourcing in a relatively short period of
  time may include reduced costs and increased productivity, there will be other
  objectives that are only achievable over the long-term. For example, an objective of outsourcing may include jointly developing complementary skills and
  capabilities in a particular process. The achievement of this objective will
  contribute to the long-term capability of the sourcing organisation to develop
  new innovative products – an overall strategic objective of many organisations.

The outsourcing objectives set may have to be refined as conditions change. For
example, although the sourcing organisation at the beginning of the relationship
established an adversarial relationship in order to attain the lowest price, the
supplier may have developed the knowledge and skills over time to become a
source of innovation and learning. In this case, the sourcing organisation may
decide to develop a more collaborative relationship in order to more fully leverage
these supplier capabilities. In fact, a major challenge in the outsourcing process is
achieving short-term cost reductions whilst allowing for the potential long-term
objective of developing a relationship that enables the innovation and learning
capabilities of the supplier to be leveraged more fully. Also, the level of experience
with outsourcing will influence the objectives set. When first approaching outsourcing, many organisations tend to outsource peripheral activities with the key
objectives of reducing costs and enhancing quality. However, as organisations
become more experienced with outsourcing and managing relationships with
suppliers, the objectives are likely to become more ambitious, encompassing the

long-term strategic aspirations of the organisation such as developing more inno-vative capabilities through the establishment of collaborative relationships with suppliers. Nevertheless, the sourcing organisation should ensure that its require-ments are not outpacing the capability of the supplier to meet them.

9.3 Supply relationship strategy

Selecting the most appropriate supply relationship will have a significant influence on the success of the outsourcing strategy. In particular, the choice of the relation-ship type must reflect the overall objectives for outsourcing the activity. The objectives established for a non-critical and critical activity will differ. For example, in outsourcing a non-critical activity the objectives are likely to be tangible and include factors such as reduced costs, better service levels in the form of on-time deliveries, inventory reduction and better quality. In this case, an adversarial relationship will be more appropriate particularly if there are a significant number of capable suppliers in the supply market. In the case of a critical activity, as well as including tangible factors, the objectives are likely to include other intangible factors such as top management commitment, joint collaboration in new product development, buyer – supplier problem solving and greater teamwork. In this case, a collaborative relationship will be more appropriate. At this stage, it is worthwhile distinguishing between collaborative relationships and the partnership philosophy. Partnership sourcing – involving equality between partners and the mutual sharing of benefits – has often been advocated as a more superior arrangement than power relationships. However, collaborative relationships with suppliers can still be pursued with suppliers even though the buyer possesses the balance of power in the relationship.

   Much of the interest in the partnering philosophy by Western companies has evolved from the experiences of the Japanese companies in the automotive indus-try. Companies such as Toyota and Honda achieved much higher levels of productivity and quality than their Western counterparts. This success was due in large part to the application of lean production methods, which involved working closely with their suppliers. However, recent doubts have been expressed about the earlier interpretation of buyer – supplier relationships in the Japanese car industry that have led to the interest in lean production and lean supply in Europe and North America. This has led to the challenging of the general applic-ability and measurable benefits of lean production and lean supply. Womack and Jones (1996) argue that lean production practices represent a universal set of principles that can achieve the same benefits outside Japan to those found in Japan itself. In particular, some have argued that the relationships pursued by the Japanese auto-assemblers were not based on equality of power in the buyer – supplier relationship. For example, Kamath and Liker (1994) have found that dependency rather than partnership best describes a number of relationships in Japan. Cox *et al.* (2000) have argued that Japanese lean production involved the lead company constructing a supply chain in which it attempted to achieve dominance over obliging suppliers. These obliging suppliers would then collaborate with the lead company to provide innovation and reduce costs. Lewis (2000) argues that the predominance of Japanese exemplars raise legitimate concerns about cultural specificity. Although, benchmarking studies have benefited from close attention to actual practice, many have largely ignored the wider economic and market conditions (Katayama and Bennett, 1996). For example, Toyota enjoyed a set of circumstances such as consistently rising demand that enabled the company to sustain a high flow through its own factories and those of their suppliers. Toyota has a unique history and geographic setting that have facilitated the practice of Just-in-time and Kanban systems.

Oliver and Hunter (1998) have questioned the competitive impact of leanness. Investigations into the relationship between profitability and lean production by Oliver and Hunter found no statistical significance between high and low users except that high users exhibited much higher volatility in profits. Burnes and New (1996) argue that partnership may not always be the only way forward for supply chain improvement. In competitive and increasingly global industries, suppliers who prosper will be the ones that can achieve world-class performance regardless of the nature of the relationship with their customers. Also, others have challenged the applicability of the methods associated with lean production to Western companies (Williams *et al.* 1994). Williams *et al.* (1992) have argued that lean production methods have not been applied universally across Japan but evolved from studies of the Toyota Production System (TPS). It must be pointed out that other Japanese companies have not adopted the principles of lean production to their fullest extent. For example, Nissan believed it was more practical and economical to keep a larger amount of inventory on hand than Toyota due to the fact that Nissan's plants were more geographically dispersed (Cusumano, 1994).

The approach proposed here in the choice of the relationship is that the sourcing organisation should do what is best for itself in terms of achieving maximum competitive advantage. In certain circumstances, the sourcing organisation can gain more by pursuing a relationship where it possesses the balance of power rather than pursuing a relationship based upon equality and the mutual sharing of benefits. The buyer can achieve the benefits of collaboration whilst still maintaining the balance of power in the relationship. In other words, in order to achieve the most beneficial supply arrangement the organisation may pursue a buyer – supplier relationship – ranging from close collaboration to adversarial – which will enable it to maximise competitive advantage. For example, the organisation may use its influence to obtain reductions in inventory and cost reductions, which in turn have a positive impact on the achievement of its own competitive position. Futhermore, this may ensure greater flexibility in that the organisation will not get locked into a long-term relationship with a supplier whose technology may become obsolete. The potential of the buyer to pursue this approach will be influenced by the importance of activity being sourced and conditions in the supply market. This approach is reflected in the following section, which outlines the potential relationship types and the appropriate conditions for each type.

9.3.1   Categorising the relationship

There are two key influences on the relationship type that should be chosen.

- *Activity importance* – the closer the activity impacts upon business success the greater the strategic importance of the activity being outsourced. The importance of the activity being outsourced is a reliable indicator of the attention that the sourcing organisation should give to managing the relationship with the supplier. Already, it has been shown how concepts such as competitive analysis, understanding the concept of customer value and the critical success factors (CSF) methodology can be used to assess the importance of the activity under scrutiny. However, there are other factors that can indicate the importance of the activity in a sourcing context. The costs associated with sourcing the activity in relation to other purchases are a significant indicator of its importance. For example, in the manufacture of a car the sourcing of the engine or gearbox is much more important than that of the headlamps or interior trim. The volume required is also an indicator of the importance of the activity. The importance of the activity will influence the level of strategic attention that should be given to managing the buyer – supplier relationship.

- *Supply market risk* – this refers to factors in the supply market that can create difficulties for the buyer in managing the relationship with the supplier. Many of these factors have already been considered under supply market risk in Chapter 8. For example, the number of suppliers in the supply market, competitive demand for supply and the relative size of other buyers are key factors that should be considered at this stage. These factors are sound indicators of supplier power. It has also been shown how risk and uncertainty in the supply market influenced the choice of sourcing option. Clearly, these factors will influence the way in which the sourcing organisation will manage the relationship with the supplier. For example, a high level of risk in the supply market will necessitate the development of a supply strategy to secure long-term supply and the employment of mechanisms to encourage the supplier to make investments that are specific to the needs of the buyer.

These two dimensions are represented on the matrix in Figure 9.1

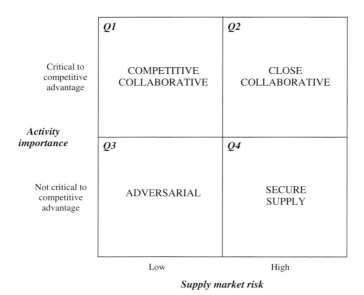

Figure 9.1 Relationship strategies

9.3.2 Critical activities

Quadrant one – competitive collaborative

In this quadrant, the sourcing organisation has two options in terms of relationship type.

Competitive collaborative – this option is most appropriate when the supply market is highly competitive with many suppliers competing for business. Consequently, suppliers have limited bargaining power and customers can easily switch suppliers. For example, this approach can be employed when the buyer is sourcing a component that is technology-intensive and rapidly changing. Rather than developing a long-term relationship, it is more prudent to manage the relationship for as long as the supplier maintains a strong position in the technology. This allows the buying organisation to switch to other competing suppliers that have become more competent in the technology as a result of technological advances. Dell has adopted a similar approach with many of its electronic components suppliers where changes in technology can rapidly erode the capabilities of certain suppliers. As a result, the relationships with these suppliers lasted as long as they maintained their leadership in technology and quality (Magretta, 1998). Moreover, due to the number of suppliers in the supply market it is possible to share the business amongst more than one supplier. This approach has the potential to reduce the risks for the sourcing organisation. In a study of both US and Japanese carmakers, Bensaou (1999) found some manufacturers use up to four capable suppliers to procure the same high-value component. For example, one such company kept three suppliers as its primary source for instrument panels and dashboards. Each of these suppliers has a promise of repeat business from the customer. However, the contract stipulates that the relationship will continue unless 'something adverse happens'. This arrangement provides the three suppliers with sufficient incentives to take a longer-term view and make the investments in R&D or process technology. The carmaker keeps these three suppliers 'on their toes' as it switches some of its annual requirements from one to another each time a problem with quality or delivery reliability is identified. This approach is employing some of the features associated with adversarial relationship including:

- the customer relying on a number of suppliers who can be played off against each other to gain price concessions and ensure continuity of supply;
- the customer allocating only limited amounts of business to suppliers to keep them in line; and
- the customer assuming an arms-length posture and using only short-term contracts.

Close collaboration – rather than have a number of suppliers competing against each other for the business, it is possible to establish a close long-term strategic collaborative relationship with one of these suppliers. The characteristics of this option are discussed in more detail in the following section.

Quadrant two – close collaboration

Close collaboration involves the adoption of a long-term collaborative relationship between the buyer and supplier. This approach is suitable in the case of a critical activity for which there are a limited number of suppliers in the supply market. Adopting this approach allows the buying organisation to establish and build a mutually advantageous relationship with the supplier. Through specialised investments in the relationship the buying organisation and the supplier can develop a relationship that delivers benefits that are unavailable to competitors. Essentially, the sourcing organisation is attempting to create causal ambiguity in the relationship with the supplier; namely obtaining benefits in the relationship that are unobtainable by its competitors. For example, the sourcing organisation and supplier may establish an agreement to jointly develop a new product or process technology. As part of this arrangement, each party may sign a confidentiality agreement that prevents the supplier from divulging any gains to competitors of the sourcing organisations. The supplier may also locate nearby in order to offer better service levels. The presence of relation-specific investments creates a high level of mutual dependency between the buying organisation and the supplier. The buying organisation has higher costs if switching to another supplier. Similarly, the supplier has higher costs if switching to a new customer because of its high dependency on the buying organisation. The characteristics of collaborative relationships are the following.

- *Long-term focus* – collaborative relations require long-term commitments from both the buyer and supplier. For example, in a manufacturing context the buyer will give the supplier the contract for the life of the component. Many of the benefits achievable from more collaborative relationships such as the greater co-ordination of inter-organisational activities and productivity improvements can only be attained over the long-term. By giving more business to the supplier, the buyer can benefit from lower prices that arise from the supplier achieving scale economies.
- *Reduced supply base* – due to the resource commitment associated with collaboration, the customer organisation must reduce the number of suppliers it does business with. It is not possible to develop close collaborative relationships with a large number of suppliers. Administrative costs can also be reduced for the buyer and supplier through the reduction of purchase orders, invoices and inspection documentation.
- *Joint problem solving* – at the heart of collaboration is a willingness on the part of the buyer and supplier to resolve any problems that affect the quality of the relationship and in turn business performance. Typical areas for joint problem-solving include cost reduction, quality improvements and lead-time reduction. Rather than switching to another supplier, the buyer should also provide assistance to the supplier in order to enhance its performance. Over the long-term, the presence of joint problem solving will lead to mutual productivity improvements and lower costs for both the buyer and supplier.
- *Information exchange* – in a collaborative relationship there are two-way information flows with the buyer and supplier co-operating on a range of issues including product design, technical capability and service levels. The communication patterns between the buyer and supplier are normally multi-functional involving a range of functions from each organisation. Effective information exchange also encourages joint problem solving between the buyer and the supplier.
- *Sharing the benefits and risks* – it is important that both the buyer and supplier share both the potential risks and benefits associated with the relationship. For example, a potential risk for the buyer when bringing a new product to market is lower sales than expected. Therefore, in a collaborative relationship, the buyer and supplier may share the costs associated with designing and bringing the new product to market in order to alleviate such risks.
- *High mutual dependence* – with both the buyer and supplier in a long-term collaborative relationship, there is a high degree of dependency between both parties. High mutual dependence reduces the risk of the supplier switching to another buyer and can create greater loyalty from the supplier.

In the car industry these types of collaborative relationships have been established for components such as power steering, suspension, and air-conditioning systems.

The nature of competition in the car industry is increasingly based upon the technology and design of these critical systems. During design and manufacture, a high level of interaction and interdependency exists between these systems and the rest of the vehicle. The creation of high mutual dependency is essential to achieving the benefits associated with collaboration. High mutual dependency between the buyer and supplier can be developed in the following areas.

Inter-organisational information systems

Information technology can strengthen collaboration in buyer – supplier relationships. For example, electronic commerce technologies can reduce the costs of integrating buyers and suppliers and through electronic networks firms can achieve an integration effect by linking processes at the buyer – supplier interface. A key enabler of collaborative buyer – supplier relations is accurate customer demand forecasting. This is an area where electronic commerce technologies have had a major impact. One industry where supplier reliability and effective forecasting is crucial is in the retail industry. For example, over-supply means the retailer is reducing cash flow, whilst under-supply means gaps on the shelves and poor customer service. Large retailers are now linking with suppliers via the Web taking advantage of the open standards technology of the Internet (Hewson, 1999). Futhermore, carmakers such as Chrysler have also used information technology to build stronger relationships with suppliers by using a computerised on-line system that transfers delivery and quality to suppliers in real time (Anonymous, 1995). These developments create a higher level of dependency in the relationship with both parties making shared investments in information technology in order to reduce costs. Information technology can enhance the customer's information management capabilities and transaction processing efficiency that in turn can be used to foster collaborative relations with suppliers.

Supplier involvement in product design and development

Greater collaboration in product design and development is another area where a high level of dependency can be created. The trend towards greater supplier involvement in product design is not surprising when one considers how dependent manufacturers are upon their suppliers. For example, purchased materials frequently account for 60% or more the total product cost of a manufacturer. In addition to the cost factor, the quality and cycle time of the manufacturer are determined by those of its key suppliers. It has been estimated that approximately 50% of a manufacturer's quality costs can be traced to purchased materials (Crosby, 1984). Suppliers are critical team members who can assist through initial design suggestions, technological contributions, and quality assurance considerations, all of which contribute to efficient manufacture and minimisation of design-to-market cycle time (Mendez and Pearson, 1994). For example, Chrysler have found that long-term collaborative relationships are essential to speeding the product development and design process (Dyer, 1996). Under its old system, Chrysler devoted 12 to 18 months of the development process to sending out bids for quotations, analysing bids, re-bidding, negotiating contract, and bringing suppliers on board and up to speed. After selecting suppliers, Chrysler would have to spend additional time responding to problems encountered when trying to manufacture a part they had not designed. Often suppliers did not even know they had won the business until 75 to 100 weeks before volume production. Under the new system, suppliers become involved at the concept stage (about 180 weeks before volume production), giving them an extra 18 to 24 months to prepare for volume production and additional time to work out potential problems early in the process. One strategy that is facilitating the effective implementation of new product development is concurrent engineering. Concurrent engineering makes use of multi-functional teams and these may be part of the infra-structural features for the supply chain, involving both internal and external representatives (Saunders, 1997). The importance of having all the internal representatives, such as research and development, engineering, purchasing, production, and logistics in the process is recognised. Concurrent engineering is also being facilitated through Internet technologies as evidenced by Caterpillar and its suppliers. Caterpillar has adopted an 'open' standards approach with its suppliers which enables partners to exchange spreadsheets, charts, documents, scheduling charts, databases and computer-generated drawings via extranet applications (Houlder, 1997).

Integrated order management and delivery systems

Many companies have been working with their suppliers to create more effective order management and delivery systems. This involves the buying organisation and the supplier collaborating to achieve improvements in areas such as inventory reduction, delivery lot-size reduction and purchase and invoice reduction. In fact, many organisations have been implementing delivery arrangements similar to the just-in-time (JIT) philosophy. Under the JIT arrangement, the buying organisation and the supplier are expected to work together to satisfy the specific needs and expectations of the buyer, to achieve better cost control, reached by long-term and exclusive agreements. The success factors associated with JIT are also the prevailing conditions for collaborative relationships, and therefore demonstrate that this relationship strategy is an appropriate framework for implementing JIT arrangements. Joint problem solving between the customer and the supplier is essential in the implementation of JIT (Yasin *et al.*, 1997). In a study of component makers in the UK automotive industry Pickernell (1997) found that the Japanese assemblers had helped their suppliers in the introduction of measures, which allowed them to cope with more frequent deliveries with minimal stocks. The Japanese assemblers brought suppliers a sense of what JIT supply should involve in terms of stability of demand. The presence of high mutual dependency between the buying organisation and the supplier is critical to the success of JIT. For example, in their study of high dependency theory in Japanese industry Oliver and Wilkinson (1992) have found that it is crucial to the success of practices such as JIT and total quality management. For example, with high inventories insulating each stage in the production process between the customer and the supplier mutual dependency is low. However, if inventory is eliminated in each stage then mutual dependency is high. High dependency is also reinforced through equity sharing and employee transferring between the Japanese companies and their suppliers.

Illustration 9.2

The development of knowledge sharing at the buyer–supplier interface

Toyota and other leading Japanese car makers such as Honda have developed knowledge-sharing routines with suppliers that have created superior inter-organisational learning. Dyer and Nobeoka (2000) have explored the knowledge-sharing routines developed by Toyota and its suppliers. Toyota and its suppliers were chosen for a number of reasons including the following.

- Toyota is the largest Japanese company and is regularly considered as the best managed and the most respected company in Japan.
- Toyota has been the most successful company in the diffusion of lean production techniques such as kanban, continuous improvement and inventory reduction in its supply chain.

Dyer and Nobeoka's analysis focused on the 'production network' – Toyota and its network of suppliers – rather than the individual firm. They show how inter-organisational collaboration can be more effective in the generation, transfer, and recombination of knowledge than that within a single organisation. This is due to the greater diversity of knowledge that exists within the network. The routines that facilitate knowledge sharing among the network members were considered. Particular emphasis was given to how the Toyota network is designed to facilitate the sharing of tacit knowledge. Toyota has created a strong network identity through the establishment of rules that support co-ordination, communication, and learning. Co-ordinating principles have been established to support co-ordination among the participating organisations. This co-ordination is achieved by creating and maintaining an identity for the network as well as mechanisms to support knowledge transfers among suppliers. In Japan, Toyota's network is known as the 'Toyota Group'

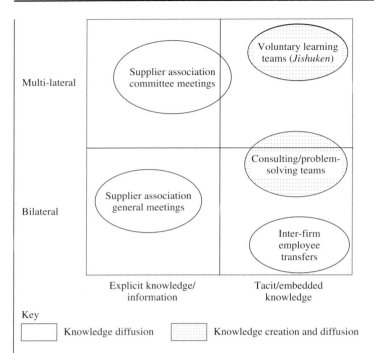

Figure 9.2 Toyota's network-level knowledge sharing processes (Dyer and Nobeoka, 2000) © John Wiley. Reproduced with permission.

and Toyota openly promotes a philosophy within the Toyota Group called 'coexistence and co-prosperity' – *kyoson kyoei* in Japanese. Toyota creates a shared network identity through the development of the following network-level knowledge acquisition, storage, and diffusion processes.

- *The supplier association* – a network-level forum for creating a shared social community, inculcating network norms, and sharing knowledge – mostly explicit.
- *Toyota's operations management consulting division* – a network-level unit responsible for knowledge acquisition, storage, and diffusion within the network.
- *Voluntary small group learning teams* – a sub-network forum for knowledge sharing that creates strong ties and a shared community amongst smaller suppliers.
- *Inter-organisational employee transfers* – this may involve engineers from suppliers working at Toyota's production facilities on an on-going basis.

Toyota established various bilateral and multilateral processes to facilitate the sharing of both explicit and tacit knowledge. Figure 9.2 illustrates the key network-level routines that facilitate the sharing of different types of knowledge within Toyota's production network.

Some of the processes are designed primarily for knowledge diffusion – supplier association meetings, Whereas other processes result in both knowledge creation and diffusion – consulting problem-solving teams. These processes match the type of knowledge with a process that is the most effective and efficient at transferring that type of knowledge. For example, the supplier association is most effective for disseminating explicit knowledge such as market trends and production levels. Alternatively, the voluntary learning groups are most effective for multilateral transfers of tacit knowledge such as processes and philosophies. However, there are a number of problems in relation to knowledge sharing within a supplier network including:

- encouraging members to participate and openly share knowledge;
- members accessing knowledge and refusing to contribute or exiting the network; and
- transferring both explicit and tacit knowledge.

Toyota addressed these problems by creating a highly inter-connected network in which members strongly identify with the 'core firm'/network. They also established clear rules for participation in the network's knowledge-sharing activities. For example, Toyota openly shares production-related knowledge with network members. Any production-related knowledge that Toyota or the supplier possesses is accessible to any member of the network because it is considered to be the property of the network. However, knowledge related to certain product designs that Toyota possesses is kept proprietary.

Source: Dyer, J. H. and Nobeoka, K. (2000). Creating and managing a high-performance knowledge-sharing network: the Toyota case. *Strategic Management Journal,* **21**, 345–367.

9.3.4   Non-critical activities

Quadrant three – adversarial

In this case, the activity can be readily sourced from a wide number of suppliers in the supply market. The strategy in this case is to source the activity from the most competitive supplier in terms of price, delivery and quality. For example, in a manufacturing context this type of relationship is appropriate for highly standardised items. These types of items require little or no customisation by the customer. They are based on a straightforward technology that requires little engineering effort and expertise from suppliers. The supply market is extremely competitive with many suppliers aggressively competing for business from customers. The straightforward nature of the activity means that the sourcing organisation has and can easily acquire sufficient knowledge to specify clearly its requirements. Furthermore, poorly performing or opportunistic suppliers can be readily replaced because of a competitive supply market. The adoption of an adversarial relationship is most appropriate in this case. An adversarial approach has the following characteristics.

- The presence of competition between suppliers in the supply market reduces the bargaining power of each supplier. This will create a constant pressure on suppliers to reduce costs. Cost reductions are obtained by having a large number of suppliers compete for a greater share of the buyer's business by offering more competitive prices.
- In a competitive supply market, suppliers will be supplying many buyers and therefore should be achieving economies of scale. These scale economies should lead to benefits for the buyer in the form of lower prices.
- An adversarial relationship is more flexible with the buyer being able to rapidly switch supply sources. Using short-term contracts, the buyer will be in a position to switch to other suppliers that improve their capabilities in relation to competing suppliers. Moreover, the costs incurred by the buyer in switching suppliers should be low enough to ensure that the buyer benefits from supplier improvements.
- In terms of managing the relationship, adversarial relationships are less resource intensive than close collaborative relationships. Normally, there is only a limited level of interaction between a number of functions in the buyer and supplier. With the buyer exerting the greater influence, the information flow is likely to be uni-directional from the buyer to the supplier.
- In fact, the flow of information is limited and based upon the parameters established in the contractual agreement. In particular, the information communicated to the supplier will include a description of the item being sourced, the required delivery dates and other information relevant to sourcing requirements.
- By keeping multiple sources of supply the buyer can prevent any uniqueness or dependency being built into the relationship with the supplier. Suppliers of similar items are regarded as homogenous. The only distinction between each supplier is the price that can be extracted through the power of the buyer.

In a study of both US and Japanese auto-makers, Bensaou (1999) found that suppliers classified in the adversarial category normally get repeat business even though short-term contracts are employed. It was also found that these relationships could be positive and collaborative without displaying some of the attributes associated with collaborative relationships such as long-term commitments. However, the buyer should be aware of the attempts of suppliers to build any dependency into the relationship. Furthermore, developments in the supply market such as mergers and acquisitions amongst suppliers can affect the balance of power in the relationship. The buyer should seek out strategies that reduce the level of attention that has to be given to managing the relationship with these suppliers. Products or services being sourced from these suppliers are not critical to the success of the buying organisation. Cleaning services, office supplies, maintenance services, and standard manufacturing components are examples of these types of products or services. For example, many manufacturers reduce the level of resource commitment to the sourcing of standard components by employing distributors to manage the supply of standard component portfolios. This arrangement allows the buyer to reduce its supply base and the amount of resource that has to be allocated to sourcing the supply of the items. Advances in information technology have also enhanced the capabilities of these distributors to include services such as electronic order management and material usage analysis.

Quadrant four – secure supply

The major difference between this quadrant and the Adversarial quadrant is the nature of the supply market. In this case, there are factors in the supply market, which can create uncertainty and vulnerability in supply. For example, the supply market may be highly concentrated with a few large suppliers. Careful attention should be given to ensuring continuity of supply. In particular, this should involve carefully monitoring the factors that create uncertainty in the supply market (Kraljic, 1983). However, there are a number of strategies that the buyer can consider in order to pre-empt any potential problems with supply.

- Establish a longer-term relationship with the supplier in order maintain supply continuity.
- Operate a consigned stock arrangement with the supplier to allow for any sudden shortages in supply. A consigned stock arrangement involves the suppliers maintaining a stock in the customer's facility and under the control of the customer.
- In a manufacturing context, it is possible to redesign the product so that the component that is subject to supply uncertainty is substituted for another more readily accessible component in the supply market. For example, in the electronics industry, designers are under constant pressures to redesign existing products to take advantage of changes and innovations in the supply market.
- The buyer can maintain some capability internally to carry out the activity in the event of supplier failure. For example, Bensaou (1999) found that Japanese car makers and US pursued this strategy for items such as bumper fascia and glass products. In the event of supplier failure, both the US and Japanese car makers would find it difficult to rapidly switch supply sources. Consequently, the car markers keep some internal manufacturing capability for the items involved.

The analysis presented has shown how the importance of the activity and the level of supply market risk influences the choice of relationship strategy. However, there are other influences that will impact the relationship strategy.

An organisation's existing set of relationships and activities already outsourced to its current suppliers are an important influence. In many circumstances the volume or monetary value of business the sourcing organisation has with a supplier may have a more significant influence on the relationship strategy chosen than the relative importance of certain activities. Furthermore, many organisations have a portfolio of activities with varying degrees of importance with their existing suppliers. Adopting an adversarial relationship strategy with a current supplier in the case of non-critical activities is not always appropriate. An organisation may already have a collaborative relationship with a supplier that it is considering outsourcing a non-critical activity. In fact, the outsourcing of a non-critical activity may be part of a wider strategy of further strengthening the relationship with this supplier. Moreover, the outsourcing of an activity may alter the nature of the current relationship with a supplier. An organisation may decide to outsource a number of other activities to a high-performing supplier in order to move a predominantly adversarial relationship towards a more collaborative basis. The current and previous performance of a supplier can be an important influence on the relationship strategy. In fact, the sourcing organisation may view the overall capabilities of the supplier as the primary influence on the relationship rather than the importance of an individual outsourced activity. However, the sourcing organisation must ensure that the supplier is delivering on the performance measures established for each outsourced activity. In fact, the ability of the supplier to meet the performance levels demanded by the buyer will significantly influence the nature and duration of the relationship.

9.4   Supplier development

Supplier development is another strategy that organisations can employ when considering outsourcing. It is particularly valuable in situations where organisations wish to outsource activities to suppliers that they are currently doing business with. The sourcing organisation can develop the capabilities of suppliers in areas, which it regards as no longer critical to its success and therefore are suitable candidates for outsourcing. Supplier development is often closely associated with collaborative buyer–supplier relationships. Supplier development involves a long-term co-operative effort between a buying firm and its suppliers to upgrade the supplier's technical, quality, delivery, and cost capabilities and to facilitate ongoing improvements (Watts and Hahn, 1993). The overall aim is to develop a mutually beneficial relationship that will help both firms compete more effectively in their respective markets. Formal supplier development programmes were first employed by the Japanese car-assemblers to develop the capabilities of their key suppliers. These companies had to work jointly with their key suppliers to achieve the required improvements in cost and quality in order to meet the challenge of increasing customer demands and intense competition. Supplier development is a major long-term strategic initiative, which strikes at the core of the competitiveness of the buying organisation. For many companies, with a considerable level of the final cost of bought-in components or services, it is extremely difficult to significantly improve quality or delivery, reduce costs or bring new products to market without a major effort from suppliers. For example, supplier development requires both the buyer and supplier to commit financial, capital, and personnel resources; to share timely and sensitive information; and to create an effective means of measuring performance (Handfield *et al.*, 2000).

Many companies are engaged in supplier development without having any formal programme. For example, joint problem solving in quality and design issues may be described as a form of supplier development. However, a formal and planned supplier development programme can deliver considerable benefits to both the buyer and supplier organisation. For example, Watts and Hahn (1993) found that some of the most important outcomes in supplier development included improvements in the quality of supplier material, supplier delivery performance, supplier service, cost reduction in incoming material, development of supplier technical capabilities and rationalising the supply base. In certain circumstances supplier development offers the buying organisation the potential to use local suppliers as opposed to sourcing from foreign overseas suppliers. Many governments are now encouraging large companies – both locally- and foreign-owned – to source much of their supply requirements from local suppliers. As part of this process governments are offering assistance and incentives to organisations to develop the capabilities of suppliers within their own regions. These companies are also finding that stronger collaborative relationships can be developed more readily with suppliers located locally than with foreign overseas suppliers. However, the decision to employ supplier development in this context will be based primarily on the potential of the supplier to develop the capabilities required for the buying organisation. For example, in peripheral regions such as Ireland many multinational manufacturers have had to source much of their supply requirements from overseas because local suppliers lack the capabilities to supply in the volume and quality required.

Supplier development is often considered to be at the heart of the partnership philosophy – permeating the entire relationship. Critical elements of supplier development such as a long-term perspective, cross-functional communication, top management involvement, cross-functional teams and a total cost focus are often associated with partnership sourcing (Krause and Ellram, 1997). In fact, many of the barriers to partnership sourcing are also applicable to supplier development including an emphasis on price, a short-term focus, and a prevailing culture of adversarial relations. Hines *et al.* (2000) have also identified a number of institutional barriers to Western companies adopting supplier development including a lack of knowledge of how to develop suppliers, an unwillingness of companies to buy-in to the process and a lack of best practice exemplars. The success of the supplier development programme will very much depend upon the buying organisation's attitude towards it supply base. For example, if the buyer has worked with its suppliers on a competitive/adversarial mode then it will only regard supplier development as another tool to squeeze further price reductions from its suppliers. Supplier development is a two-way process with both the buyer and supplier sharing the associated benefits and risk. In fact, supplier development can be considered to as a means of identifying and developing new suppliers where no adequate ones exist. Therefore, supplier development is another strategy that should be considered in the context of outsourcing.

9.4.1   Integrating supplier development into the outsourcing process

Stage 1   identify potential activities
In this case, the impetus for considering supplier development emanates from outsourcing evaluation. Supplier development should be considered in outsourcing evaluation in the following situations.

- In the case of a non-critical activity in which the buying organisation is more competent than external suppliers, the buying organisation can consider developing the capabilities of a supplier in this area rather than continue to undertake the activity internally.
- In a case where the company possess leadership in a technology that is diminishing in importance in its product markets, developing the capabilities of the supplier in the technology is a potential option.

The objectives of the supplier development initiative should reflect the objectives for outsourcing the activity. The importance of the activity being outsourced will determine the amount of resource required for supplier development and extent of the improvements required from the supplier. For example, the supplier development programme may involve the buyer offering assistance to the supplier in order to reduce costs, improve quality and improve delivery performance in one of its product lines. Alternatively, the buyer may adopt a supplier development programme to teach suppliers in a range of processes that will achieve dramatic improvements across all areas of the supplier's business. Also at this stage, consideration must be given to the feasibility of pursuing supplier development in terms of the resource requirements and potential likelihood of success. Supplier development can be extremely resource intensive and should only be pursued with a very limited number of suppliers.

Stage 2 identify potential suppliers

Potential suppliers can be identified from the evaluation of the capabilities of the buying organisation in relation to external suppliers in outsourcing evaluation. In fact, the analysis of supplier capabilities will assist in determining the areas in which the supplier's performance is lacking and serve as a useful starting point for determining how the development initiative should be progressed. For example, the analysis may have revealed deficiencies in areas such as quality, delivery times, cost control or cycle times. The choice of supplier will also be influenced by the future plans of the buying organisation for its supply base. For example, the buying organisation may see potential in outsourcing an activity to increase the amount of business it has with a particular supplier. Although the supplier may not be attaining the levels of performance required in the activity, a supplier development programme offers a means of raising the performance levels of the supplier. Such a strategy will enable the buyer to strengthen the relationship with this supplier and create a higher level of dependency in the relationship. Moreover, due to the costs associated with searching for and evaluating a new supplier, working with a current supplier to improve performance is a more beneficial option.

Stage 3 obtain supplier commitment

Hartley and Choi (1996) have found that gaining commitment from supplier top management is the most critical factor in supplier development. The buying organisation may meet with representatives of the supplier in order to determine the causes of problems and the required corrective actions. If the supplier fails to demonstrate the necessary commitment to the proposed supplier development initiative then an alternative source of supplier should be considered. In order to demonstrate commitment, suppliers must be willing to admit to performance weaknesses in particular areas of their operation. The supplier development initiative will not work unless the supplier recognises the need for improvement. The supplier may also be uncomfortable with opening up their business to scrutiny from the buyer. For example, during the supplier development initiative the buyer will gain considerable knowledge of supplier operations – both positive and negative aspects. However, the buyer must convince the supplier of the benefits of the supplier development initiative and demonstrate its commitment through increased business and a longer-term relationship with the supplier. The buyer must also ensure that it gains the respect of the supplier through addressing any problems with its own operations.

Stage 4 establish buyer – supplier development teams

Supplier development teams should be established both within the buyer and supplier organisations. The top management of both the buyer and supplier should be involved in determining the composition of their respective teams and commit the necessary resource to demonstrate their commitment to the supplier development initiative. Cross-functional participation from both the buyer and supplier is required due to the potential range of problems that have to be addressed. In the case of the buyer organisation, the team should be composed of people with specialist expertise from areas such as quality, manufacturing and design. As well as being skilled in their area of expertise, buyer team members should have the skills to train employees and facilitate changes within the supplier organisation. Team leaders within the both the buyer and supplier organisations should be chosen. Hartley and Choi (1996) argue that the supplier team leader must have credibility across the supplier business and be in a position to persuade supplier personnel of the benefits of the changes that will result from the supplier development initiative. The team must ensure that any changes in supplier operations as a result of the supplier development initiative are sustained after the assistance by the customer is complete. It is important that there is interaction between both teams on an ongoing basis in order to both demonstrate commitment to the supplier development effort and strengthen the buyer – supplier relationship. In fact, the buyer may find that many supplier-related problems may have arisen due to poor communication and a lack of openness at the buyer – supplier interface.

Stage 5   define terms of engagement

It is important to have a detailed understanding of supplier operations. Knowledge of supplier capabilities will inform this part of the analysis for the buyer. Potential sources of information on supplier capabilities include the buyer's supplier evaluation system as well as the supplier's own internal performance monitoring systems. Having a full understanding of supplier operations will assist in identifying problem areas and causes of poor performance. Once the problem areas have been identified, the buyer and supplier must agree on the performance measures that will be used to assess the success of the supplier development initiative. Potential measures may include percentage of quality improvements to be achieved, productivity improvements, cycle time improvements, percentage of delivery improvements, and percent of cost savings to be shared. Agreement must be made on how the benefits of the supplier development initiative are shared. For example, in cases where technology development is involved non-disclosure agreements will have to be drawn up. The buyer and supplier must agree to time-scales for the achievement of the performance measures identified. This will also involve agreeing upon the role of the buyer and supplier in meeting the performance measures set. Suppliers must be aware of the considerable resource required in the form of people, equipment and finance to pursue the improvements required. Based upon a study of a range of companies engaged in supplier development, Handfield *et al.* (2000) have suggested the following actions to address this particular problem.

- *Initial improvements should be kept simple* – quickly discovering and correcting straightforward problems can deliver significant benefits;

- *Draw on the resource of the buyer* – for example, the buyer may provide expertise in integrating the information systems of the buyer and supplier; and
- *Offer personnel support* – rather than offer financial support, the buyer can provide the services of some of its employees to assist in the relevant problem area.

Stage 6   implement, evaluate, and sustain

This stage is concerned with implementing and evaluating the improvements as agreed in the previous stage. Evaluation of supplier performance is crucial at this stage and can be used to guide the supplier development efforts. For example, Varity Perkins places its supplier evaluation system at the heart of its supplier development efforts. Varity Perkins expects its suppliers to recognise the potential benefits of supplier development by encouraging their suppliers to compare their performance relative to their competitors. Evaluation of supplier performance will also allow the buyer to recognise any improvements that the supplier achieves as a result of supplier development. Typically, suppliers can be rewarded through *certified* or *preferred* status. Krause and Ellram (1997) argue that supplier recognition is a critical element of supplier development. For example, achieving award status from the buyer can motivate suppliers to improve particularly if this can lead to other benefits such as attracting new business from other customers. Supplier recognition can also be employed to represent the performance targets agreed for the supplier development effort. One of the greatest challenges is to sustain the associated improvements once the supplier development initiative is complete. Hartley and Choi (1996) found four factors essential to sustaining and developing the improvement throughout the supplier organisation:

- hands-on training of supplier team members;
- follow-up and measurement by the customer on an ongoing basis;
- fit of the approach with the supplier's corporate culture; and
- The support structure in the supplier's organisation.

It is important that development efforts are sustained through the development of future supplier capabilities. The ability to achieve ongoing improvements must become an integral part of supplier operations.

Illustration 9.3

Supplier development at Honda

In the 1980s and 1990s, Japanese carmakers including Toyota, Honda and Nissan established manufacturing facilities in the US. When these companies first established assembly plants in the US, they brought many of their Japanese suppliers with them. However, in exchange for being permitted to enter the US market, the US government stipulated that they should use local supply sources as much as possible. Therefore, many of these companies rapidly disseminated their innovative manufacturing practices such as JIT delivery and supplier development, which had been so successful in Japan, with US suppliers. One company in particular, Honda, placed supplier development at the heart of its relationships with its suppliers in the US. Honda's success has been very much built on its manufacturing philosophy. The success of this philosophy has depended upon both the people who manufacture the product and the suppliers that provide the parts and raw materials. Honda has established long-term, mutually beneficial relationships that involve committing the necessary effort and resources to help suppliers reach desired performance. Interaction between Honda and its suppliers is continuous with considerable movement of employees between plants. For example, Honda's 800 associates who work with suppliers often relocate to suppliers' plants for months. Supplier development has supported Honda's strategy of 'localising' supply in North America. Supplier development at Honda of America has a number of critical elements.

- *Support from top management* – top management at Honda view suppliers as critical to the success of their business. Honda identifies and develops world-class suppliers that will adopt its systematic approach to reducing costs, increasing quality and developing leading-edge technology. Purchasing is also considered as a critical function by top management because of the emphasis on suppliers. In fact, many of Honda's senior directors have been involved in purchasing as some stage in their careers.

- *Cost reduction* – target costs are set for most component parts. Honda gives suppliers assistance in meeting these targets. As part of this process, Honda breaks down costs to the component level. Suppliers are then requested to provide detailed cost breakdowns including raw materials, labour, tooling, packaging, delivery, and administration. Honda jointly develops cost tables with suppliers and uses them to compare differences across all cost elements. By comparing the cost breakdowns, Honda suggests potential improvements to the supplier in order to reduce costs. This approach has resulted in on-going productivity increases and cost reductions, which are normally shared with the supplier. As well as reducing costs for the component under analysis, suppliers are also learning how to carry out the process for application in the future.

- *Teaching self-reliance* – Honda's philosophy with supplier development is based on the old axiom – 'if you give a man a fish, you feed him for a day, if you teach a man to fish, you feed him for a lifetime'. The ultimate aim of their supplier development efforts is for suppliers to reach a level of understanding of Honda's manufacturing philosophy where they no longer require assistance. In fact, some of the suppliers have become so proficient in Honda's practices that they are teaching their own suppliers.

- *Quality improvement* – purchasing and quality aggressively tackle quality problems. Again, suppliers are given assistance where appropriate. For example, in one case, one of its small plastics suppliers did not have the capacity to produce in the required volume, which led to the quality of output deteriorating. Honda sent four people to the supplier for 10 months – at no charge to the supplier. Additional services were offered as required. As a result, the supplier achieved the necessary improvements and became a well-established Honda supplier. No incoming inspection of parts is conducted. Suppliers are constantly rated by purchasing. Top-performing suppliers in a variety of categories receive awards annually. This approach has led to dramatic improvements in quality-related problems since it first began manufacturing cars and motorcycles in the US.

- *Product research and development* – suppliers become involved in new model development years before the product enters the manufacturing stage. Suppliers normally become involved 2 to 3 years before a new model introduction. Some of its key suppliers spend up to three months working with Honda's design engineers in Japan. Honda's Marysville facilities can house up to 30 supplier representatives at once to work with Honda's engineers. The level of involvement of suppliers depends upon the importance of the part. Functional parts or systems such as air-bag modules, brakes, steering gears, alternators rely heavily on supplier technology, whereas fabricated parts such as body stampings, engine castings, interior trim pieces are less driven by technology. However, Honda still requests these suppliers to examine parts manufacturing processes and continually strive to reduce production costs. Honda encourages suppliers to question old designs and constantly challenge conventional design thinking.

Source: Fitzgerald, K. R. (1995). For Supplier Development – Honda Wins!, *Purchasing*, 21 September, 32–40.

REFERENCES

Anonymous (1995). Chrysler pushes quality down the supply chain. *Purchasing*, 13 July, 125–8.

Bensaou, M. (1999). Portfolios of buyer – supplier relationships. *Sloan Management Review*, **39**, No. 4, 35–44.

Burnes, B. and New, S. (1996). Understanding supply chain improvement. *European Journal of Purchasing and Materials Management*, **2**, No. 1, 21–30.

Cox, A., Sanderson, J. and Watson, G. (2000). Wielding influence. *Supply Management*, 6 April, 30–3.

Crosby, P. (1984). *Quality without tears*. New York: McGraw-Hill.

Cusumano, M. A. (1994). The limits of 'lean'. *Sloan Management Review*, **35**, No. 4, 27–32.

Dyer, J. H., (1996) How Chrysler created an American keiretsu. *Harvard Business Review*, **74**, No. 4, 42–56.

Dyer, J. H. and Nobeoka, K. (2000). Creating and managing a high performance knowledge-sharing network: the Toyota case. *The Strategic Management Journal*, **21**(3), 345–67.

Hartley, J. L. and Choi, T. Y. (1996). Supplier development: customers as a catalyst for change. *Business Horizons*, July–August, 37–44.

Handfield, R. B., Krause, D. R., Scannell, T. V. and Monczka, R. M. (2000). Avoid the pitfalls in supplier development. *Sloan Management Review*, **41**, No. 2, 37–48.

Hewson, D. (1999). *Teamwork with Suppliers*. Sunday Times: E-Commerce Supplement: London, 26 March, p. 8.

Hines, P., James, R. and Jones, O. (2000). *Supplier Development in Value Stream Management: Strategy and Excellence in the Supply Chain*. London: Financial Times: Prentice Hall.

Houlder, V. (1997). Design and production benefit from wired collaboration. *Financial Times: Information Technology: Make it on the Internet: Online Manufacturing*, 16 July, 34.

Kamath, R. and Liker, J. (1994). A second look at Japanese product development. *Harvard Business Review*, **74**, No. 6, 154–70.

Katayama, H. and Bennett, D. (1996). Lean production in a changing competitive world: a Japanese perspective. *International Journal of Production and Operations Management*, **16**, No. 2, 8–23.

Kraljic P. (1983). Purchasing must become supply chain management. *Harvard Business Review*, **61**, No. 5, 109–17.

Krause, D. R. and Ellram, L. M. (1997). Critical elements of supplier development: the buying-firm perspective. *European Journal of Purchasing and Supply Management*, **3**, No. 1, 21–31.

Lewis, M. A. (2000). Lean production and sustainable competitive advantage. *International Journal of Production and Operations Management*, **28**, No. 8, 959–78.

Magretta, J. (1998). The power of virtual integration: an interview with Dell Computer's Michael Dell. *Harvard Business Review*, **76**, No. 2, 73–84.

Mendez, E. G. and Pearson, J. N. (1994). Purchasing's role in product development: the case for time-based strategies. *International Journal of Purchasing and Materials Management*, **30**, No. 4, 3–12.

Oliver, N. and Hunter, G. (1998). *The Financial Impact of 'Japanese' Manufacturing Methods, Manufacturing in Transition*. London: Routledge.

Oliver, N. and Wilkinson, B. (1992). *The Japanisation of British Industry*. London: Basil Blackwell.

Pickernell, D. (1997). Less pain but what gain?: a comparison of the effectiveness and effects of Japanese and non-Japanese car assemblers' buyer – supplier relations in the UK automotive industry. *OMEGA*, **25**, No. 4, 377–95.

Saunders, M. (1997). *Strategic Purchasing and Supply Chain Management*. London: Pitman Publishing.

Watts, C. A. and Hahn, C. K. (1993). Supplier development: an empirical analysis. *International Journal of Purchasing and Materials Management*, **29**, No. 1, 11–17.

Williams, K., Haslam, C., Williams, J., Cutler, T. and Johal, S. (1994). De-constructing car assembler productivity. *International Journal of Production Economics*, **34**, 253–65.

Williams, K., Haslam, C., Williams, J., Cutler, T., Adcroft, A. and Johal, S. (1992). Against lean production. *Economy and Society*, **21**, No. 3, 321–54.

Womack, J. P. and Jones, D. T. (1996). *Lean Thinking*. New York: Simon and Schuster.

Yasin, M. M., Small, M. and Wafa, M. A. (1997). An empirical investigation of JIT effectiveness: an organisational perspective. *OMEGA*, **25**, No. 4, 461–71.

10 Establish, manage and evaluate the relationship

10.1 Introduction

Many organisations underestimate the effort involved in implementing an outsourcing strategy. This lack of awareness can range from the potential pitfalls that can arise to the level of resource involved in managing the outsourcing process. For example, there is a common perception that once an activity is outsourced that responsibility for the outsourced activity rests solely with the supplier. Many organisations also lack the skills to implement an effective outsourcing strategy. Managing an external supplier requires a different set of skills than those associated with managing an internal business function. For example, the manager responsible for managing the activity when it was performed in-house may not have the range of skills or experience to manage an external provider of the activity. This lack of awareness of the effort involved in managing external relationships can be particularly serious in the outsourcing of critical activities. Often the buyer will adopt a collaborative relationship with the supplier for a critical activity. However, developing collaborative relationships across organisational boundaries presents organisations with considerable challenges from an organisation change perspective. For example, when developing collaborative relationships, a culture of confrontation can stifle the intentions of the buyer in its attempts to develop a close collaborative relationship with suppliers. In pursuing collaborative relationships, the buyer has to consider a number of issues including the adoption of an integrated approach to the management of strategic change, the pivotal role of senior managers as facilitators of change, and the involvement of those most affected by the change.

Many organisations do not give sufficient attention to analysing the capabilities of suppliers to deliver the product or service to the required standards. Outsourcing often fails because suppliers cannot meet the service levels required by the buyer. The ambitions of the organisation in its outsourcing strategy will influence the level of attention it should give to analysing supplier capabilities and managing the relationship. For example, if outsourcing has been placed at the heart of the overall strategy of the organisation and perceived as making a contribution to competitive differentiation, then the management of the outsourcing process should receive considerable attention. As well as considering the capability of the supplier to deliver the product or service, the buyer must be aware of the likelihood of the supplier negotiating and establishing a relationship primarily for the supplier's own benefit. Mistakenly, many organisations believe they can establish a relationship in which they possess the balance of power and employ adversarial mechanisms to keep suppliers in line. However, organisations are often held to ransom by suppliers because they do not possess the necessary skills to negotiate and draw up a robust contract. Opportunistic suppliers are quite adept at negotiating contracts that create high dependency and high switching costs for the buyer. Many organisations are unaware that often the critical skill of suppliers in outsourcing markets is that of negotiating the contract for their own benefit.

This chapter sets out the stages that have to be addressed when implementing an outsourcing strategy. Selecting the supplier is the first step that should be carried out. The buyer must ensure that the supplier chosen has the capabilities required to meet its needs. Once the supplier is selected, the next stage is concerned with dealing with the contracting issues. This stage will address issues such as the service level agreement, transfer of staff and assets, price and payment terms, liability, contract termination and flexbility. Once the contract is negotiated and drawn up, the buyer then establishes and manages the relationship with the supplier. The people responsible for managing the relationship must have the necessary skills and experience to ensure the relationship meets the objectives of the outsourcing strategy. Evaluation of relationship performance will be carried out on an ongoing basis addressing a number of issues including supplier performance, strength of the relationship and the level of dependency on the supplier. This evaluation serves as a framework for action, which can involve maintaining the relationship at its current level, further developing the relationship or discontinuing or reducing the scope of the relationship.

10.2   Supplier selection

One of the most important activities in the outsourcing process is that of supplier selection. Organisations must have a supplier selection strategy, which enables them to achieve the objectives of the outsourcing strategy. Supplier selection has become increasingly important as organisations outsource more activities and develop more collaborative and longer-term relationships with key suppliers. The supplier selection process can be extremely time consuming and involves considerable analysis of potential suppliers. Even in the case of selecting a supplier for an adversarial relationship, the analysis extends to considering more factors than price alone. Factors such as the relative importance of quality and service have become increasingly prominent in the selection decision. Although price is still the dominant selection criterion, the overall increase in the importance of factors such as quality and service is indicative of the changes occurring in the relationships between many buyers and their suppliers. Whereas, in the past, the nature of buyer–supplier relationships have been predominantly adversarial in nature, collaborative relationships have become more prevalent, as organisations have outsourced more critical organisational activities. Clearly, the greater the importance of the activity the more attention should be given to the supplier selection process. The objectives of the outsourcing strategy will influence the level of attention that the organisation will give to the supplier selection process. For example, in the case of outsourcing critical business activities, supplier selection is a strategic decision that will involve input from senior management. There are a number of steps in the supplier selection process.

10.2.1   Determine requirements

The starting point in the supplier selection decision involves the buyer determining what is required from the supplier. This will involve dissecting the activity involved in order to determine the skills and resources required to perform the activity. This will serve as the basis for the request for proposal (RFP), which consists of a purchase description of the product or service, information on quantities, service levels, special terms and conditions, and standard terms and conditions. The objectives of the outsourcing strategy will also assist in this phase. At the most basic level, this will include a specification of the product and service the supplier should provide to the buyer. For example, in the case of a competitive bidding situation, the specification will set out the buyer requirements in the tender documentation. In this case, the buyer's requirements from the supplier are primarily the lowest price along with clearly defined levels of service. However, in the case of a critical activity the buyer has to consider much more than the supplier's capability to deliver at the most competitive price. As well as being able to deliver the product or service, the supplier must be willing to develop a close collaborative relationship that involves cross-functional interaction and joint problem solving. This will involve the supplier agreeing to commit the necessary resource to the relationship over the long term.

10.2.2   Determine criteria for evaluation

Once the sourcing organisation has decided what is required from the supplier, it can then determine the criteria it wishes to employ to evaluate potential suppliers. A comparison of the features of adversarial and collaborative relationships can provide valuable insights into the type of criteria that should be defined for evaluation in the supplier selection process. The dominant features of adversarial relationships include an emphasis on price followed by quality, delivery and service. These features are easily quantifiable and will dominate the criteria used in the evaluation of potential suppliers. In contrast, the key features of collaborative relationships include cross-functional buyer–supplier interaction, joint problem solving, information sharing and a long-term perspective. Therefore, the factors that will be used in the evaluation are comparable to those associated with the partnership model. In this situation, as well as considering the quantitative criteria such as price, quality, delivery and service, a new set of supplier selection criteria come into consideration. Ellram (1990) uses the term 'soft' factors to describe this set of criteria. These soft factors can include issues such as top management compatibility, design capabilities, company culture and the strategic direction of the supplier firm. These factors are unique due to the collaborative nature of the buyer–supplier relationship and are similar to those identified by Ellram (1990) for a partnership relationship as shown in Table 10.1. Each of these factors may have varying levels of importance depending upon the characteristics of the activity being sourced and the nature of the supply market. Although criteria such as top management compatibility and cultural alignment are highly subjective, they are important determinants in the success of close collaborative relationships.

Table 10.1 *Supplier partnership selection criteria (Ellram, 1990, Table III, p. 12)*

Category	Criteria
Financial issues	• Economic performance
	• Financial stability
Organisational culture and strategy issues	• Feeling of trust
	• Management attitude/outlook for future
	• Strategic fit
	• Top management compatibility
	• Compatibility across levels and functions of buyer and supplier firms
	• Supplier's organisational structure and personnel
Technology issues	• Assessment of current manufacturing facilities/capabilities
	• Assessment of future manufacturing capabilities
	• Supplier's design capabilities
	• Supplier's speed in development
Other factors	• Safety record of the supplier
	• Business references
	• Supplier's customer base

From the Institute for Supply Management[TM]. Reproduced with permission.

An important consideration in the context of collaborative relationships is a long-term focus. As well as considering the current performance and capability of a supplier, the future direction and capabilities of a supplier should be assessed. The attitudes and culture of the supplier will provide a sound indication of its aspirations for the future. Moreover, even though a supplier may be deficient in an area, there may be the potential for the buyer to employ supplier development because the culture of the supplier indicates a willingness to improve and strengthen the relationship with the buyer. The financial standing of the supplier will also be a sound indicator of the ability of the supplier to deliver over the long term. The commitment of the supplier to its areas of strength should also be considered through an analysis of its current commitments and future strategy.

10.2.3 Evaluation

This phase involves evaluating potential suppliers against the criteria identified in the previous phase. The outsourcing strategy will fail if the supplier cannot meet the criteria identified. Blumberg (1998) has found that outsourcing fails because the supplier lacks the depth and breadth of service capability or cannot provide the full range of services required to support all the components of the activity being outsourced. In effect, the outsourcing process fails because suppliers cannot deliver on the requirements of the buyer. In the case of standardised items, the evaluation will focus primarily on the quantitative criteria identified. The level of evaluation will be influenced by the buyer's previous experience with suppliers identified. For example, the decision will be informed by the historical performance of the supplier. The depth of evaluation will be determined by how familiar the buyer is with the supplier in question. In many cases, the buyer will normally develop a collaborative relationship with a supplier which it has already done business with. In fact, in-depth knowledge of the supplier's capabilities and performance may have created the initial impetus for outsourcing an activity to the supplier. However, the buyer will still have to give careful consideration to the selection decision.

In relation to the outsourcing of critical activities, the buyer will have to give significant attention to evaluating a supplier for a close collaborative relationship. This will involve a thorough review of potential suppliers by the buyer, usually involving visits to supplier facilities, interviews, presentations, and information sharing. This interaction will occur at both the operational and strategic level. Top management from the buyer and supplier must be involved in communicating the expectations of each party from the relationship. The involvement of top management indicates the importance of the decision to employees at lower levels in both organisations. Interaction at the operational level is also important in the evaluation phase. The success of the relationship will be largely determined at the operational level where the relationship is managed on a day-to-day basis. In effect, this analysis is concerned with evaluating the potential fit between the buyer's and supplier's management philosophy and commitment to the same values of improvements over time.

10.2.4 Selection

In the final selection decision, each supplier will be evaluated against their ability to meet the criteria identified. If the buyer cannot select a suitable supplier that meets the criteria, then this will clearly have significant implications for the outsourcing strategy. In fact, the failure to select a supplier may force the company to abandon the outsourcing of the activity and continue to perform the activity internally. Therefore, it is important that the buyer should identify a number of suitable suppliers at a very early stage in outsourcing evaluation in order to avoid this scenario. In fact, the analysis of the capability of suppliers in relation to the internal capabilities of the buyer in the activity concerned should reveal a number of potential suppliers.

10.3 Contracting issues

The importance of establishing clear objectives for outsourcing has already been emphasised. These outsourcing objectives can also be used as a basis for drawing up a contract for the outsourcing process. DiRomualdo and Gurbaxani (1998) have found that poor results from outsourcing are as a result of a failure to carry out the following:

- define clearly specific objectives for outsourcing;
- align the contract and relationship with the strategic objectives set;
- make contracts flexible enough to adjust to changes in the business or technology; and
- ensure that the supplier has the capabilities required to meet the strategic objectives for outsourcing.

However, clearly established objectives and a well-designed contract can compensate for many of these potential pitfalls. A well-designed contract can allow for most future contingencies and how these contingencies should be dealt with. It is better to have a clear idea of what is required in order to avoid any potential gaps in the contract. The type of activity being sourced and the level of risk in the supply market will influence the design of the contract. For example, in the case of a straightforward activity with a significant number of suppliers in the supply market, the objectives of outsourcing the activity and requirements of the supplier are well defined. Alternatively, if there is uncertainty about the requirements the contract must be designed to reflect this uncertainty. The performance measures incorporated into the contract must reflect the objectives of the outsourcing strategy. For example, it is unlikely that a contract which focuses primarily on securing price reductions from the supplier will lead to the development of a relationship which is designed to foster and encourage inter-firm innovation. Organisations considering outsourcing must have an understanding of the complex business and legal issues associated with outsourcing and how some of these issues can be dealt with in a contract.

Illustration 10.1

Effective contracting

Jerome Barthelemy carried out an in-depth study of the experiences of 91 European and US-based organisations with outsourcing. The research focused on a range of business activities including information technology, telecommunications, logistics and finance. In particular, the research revealed some interesting insights into the experiences of these organizations in relation to the drawing up of contracts. For example, one European bank that had outsourced its telecommunications network had as a result experienced increased costs and poor service quality because they had rushed into the relationship with the service provider. The bank had failed to spend sufficient time drawing up the contract, which had led to a number of critical errors being made. For example the bank had to pay extra fees for basic services because the contract was not sufficiently precise. Moreover, there were no objective performance measurement clauses established at the outset. The nature of the contract had made it extremely difficult for the bank to switch service providers once the relationship had deteriorated. Some of the literature on outsourcing has popularised the notion that contracts play a minor role in collaborative relationships. However, Barthelemy argues that a good contract is essential because it establishes clearly the balance of power between the sourcing organization and the supplier. Drawing up a robust contract is important because it allows each party to the relationship to set expectations and commit themselves to short-term objectives. It also provides a safety mechanism in the event of a deteriorating relationship. From this study, Barthelemy (2003) has identified a number of key characteristics of a robust contract.

- *Precise* – cost and performance requirements by the sourcing organisation should be established and clearly specified. One of the key reasons why organisations fail to achieve the desired benefits of outsourcing is that they have limited control over supplier performance.

- *Complete* – a more complete and detailed contract will limit the potential of opportunism from the supplier and futhermore, if the contract is more complete then there is less likely to be costly renegotiations.
- *Incentive-based* – the contract should be structured to obtain the desired performance levels from the supplier and develop a healthy relationship. For example, the sourcing organisation and the supplier will share any benefits that arise from developing any innovations that are specific to the relationship. An important part of this process is ensuring that both the sourcing organisation and supplier have complementary and shared objectives. The contract should also address how the relationship will evolve over its life cycle. For example, unit-base pricing may be employed at the beginning of the relationship whilst cost-plus pricing may be employed as the relationship develops.
- *Balanced* – a contract should be balanced in protecting the interests of both the sourcing organisation and the supplier. Attempts at employing clauses to obtain benefits for only one side of the relationship will be corrosive with both parties losing. Suppliers may attempt to conceal clauses in the contract that create hidden costs for the sourcing organisation. For example, suppliers may charge exorbitant fees for services that the sourcing organisation assumes are already included in the contract.
- *Flexible* – medium and long-term contracts should be written to allow for changes in technology and the prevailing business conditions. Flexibility clauses should be written into the contract for both parties to accommodate changes in the business environment.

Source: Barthelemy, J. (2003). The seven deadly sins of outsourcing. *The Academy of Management Executive*, **17**, No.2, 87–98.

An outsourcing contract will include some of the following aspects.

10.3.1   Service level agreement

A service level agreement (SLA) is an agreement between the customer and the supplier that quantifies the service levels required by the customer (Hiles, 1994). The service level agreement should describe, the types, scope, and nature of all the services required, the times when these services should be available, and the level of performance required. Clearly defined services and service levels are critical elements of an outsourcing contract. Larson (1998) argues that the key to successful outsourcing involves defining services and service levels that can be measured and managed; can be audited; can be provided at an economic price; and give maximum value to the users of the services. The SLA should also allow the buying organisation to measure the performance of the supplier through a number of mechanisms including regular progress meetings, inspection procedures, etc. The types of performance measures employed will depend upon the outsourcing context. For example, in an information technology context Larson (1998) has identified the following as dimensions of performance.

- *availability* – identifies the proportion of the time that the service is actually accessible and usable over a defined time period;
- *reliability* – defines the frequency with which the service is withdrawn or fails over a defined time period;
- *serviceability* – measures the length of available time lost between the point of service failure and service reinstatement;
- *response* – measured as turnaround time or transfer time, for example in the case of a help desk call; and
- *user satisfaction* – related to perceived performance versus expectation.

The management of the SLA can involve a considerable resource commitment on the part of both the buyer and supplier. The management of SLAs will involve the following three parties.

- *The relationship promoter* – will be involved primarily on the measurement and examination of the performance of the supplier against the agreed service levels. The relationship promoter along with the internal customers should create an environment in which the agreed service levels are met. The relationship promoter is also responsible for implementing contract variations and dealing with charges and compensation (Larson, 1998).
- *The supplier* – is responsible for delivering the product/service, resolving operational problems and managing routine changes in accordance with the contract. The buyer–supplier interface at the operational level between the supplier and buyer focuses on the demand and supply of the product/service.
- *The internal customers in the buyer* – are responsible for integrating the product/service into the buyer's value chain. The internal customers should also be involved in determining the requirements set out in the SLA.

There are potential problems with SLAs if they are not implemented properly. The major reasons for the failure of SLAs include poor measurements, inadequate definitions, and a lack of the required resource commitment. The SLA requires a significant resource commitment on the part of the customer. For example, the customer has to agree to support the objectives of the SLA through the provision of accurate forecasts, attainable deadlines and realistic cost targets. The nature of the buyer–supplier relationship will influence the level of resource that should be committed to managing the SLA. Due to the contractual nature of the SLA it is important to employ it in the proper context. For example, in the context of an adversarial buyer–supplier relationship, contractual provisions can be included to penalise the supplier financially if at any time the supplier fails to deliver the level of service required. Alternatively, in the case of a close collaborative relationship the rigid adherence to the contract can damage a relationship, that is attempting to foster innovation and joint improvement. In this case, it is more appropriate to use the SLA as a guide in creating a service management ethos and aligning the relationship to the objectives of outsourcing. Evaluation of supplier performance should also focus more on intangible factors that contribute to the improvement of the quality of service and the health of the customer–supplier relationship.

10.3.2   Transfer of staff

Outsourcing contracts often involve the transfer of staff from the customer organisation to the supplier. As part of outsourcing an activity, the supplier may agree to the transfer of all the employees from the customer organisation involved in that activity. In the European Union, employee regulations, such as the 'Transfer of Undertakings (Protection of Employment)' Act (TUPE) require procedures to be completed before staff can be transferred to the supplier. The major effect of this legislation is to ensure continuity of employment for staff. As part of this transfer their existing terms of services must be guaranteed. In many cases, the legislation can further complicate the process and in some cases there is doubt on whether it applies. For example, in the case of the complete transfer of an in-house function to supplier, the TUPE regulations are likely to apply. However, there is a lack of clarity on whether the regulations will apply for outsourcing arrangements, which apply to only part of an organisational function (Lee, 1996). Furthermore, the TUPE regulations do not provide sufficient detail on the pension entitlements of transferred employees. The situation is further exacerbated in that such legislation is subject to frequent change. The sourcing organisation may have to employ specialist expertise to provide guidance in this area.

10.3.3   Asset transfer

A common feature of outsourcing is the transfer of assets from the sourcing organisation to the supplier. For example, the customer may transfer the in-house equipment that it used to manufacture the equipment to the supplier. This asset transfer may be dealt with through a sale agreement in which the assets are formally transferred to the supplier. In some cases, the transfer of assets may be taxed in the form of either VAT or stamp duty.

10.3.4   Price and payment terms

The price and payment terms must be agreed in the contract. This will involve agreeing when, how and to whom the payments should be made along with the amounts and structure of the payments. A mechanism should be included to allow for any increases or decreases in costs to be built into the contract. For example, in the sourcing of high technology items it is necessary to include in the contract a mechanism for negotiating price and payment terms on an ongoing basis to allow for any advances in technology. It is also important to include a formula to calculate payments for the provision of any additional services that were not anticipated at the outset.

10.3.5   Liability

There is the potential for litigation in outsourcing if the supplier fails to meet the required service levels. However, litigation is normally a last resort because it is extremely time consuming with no guarantee of success. An alternative is to write express warranty into the agreement for the supplier to indemnify the company for, any losses, costs, and liabilities associated with the supplier breaching the contract (Lee, 1996). For example, the customer organisation may suffer severe disruption to its operations and a loss of business due to supplier failure. It is therefore important to ensure that these losses are recoverable by explicitly providing for them in the contract.

10.3.6   Contract termination

There will normally be a number of conditions under which the contract should be terminated. These conditions can include the bankruptcy of either party, the refusal of either party to abide by the terms set out in the contract or non-payment by the sourcing organisation. There has to be clear agreement on these conditions when the contract is drawn up. It is important to establish at the outset that there is a minimum of disruption to operations in order to extricate either party from the contract. A reasonable time should also be given to either party to seek to rectify any problems that may arise in the relationship.

10.3.7   Flexibility

A robust contract can be a significant determinant on the success of the outsourcing process. Contracts are an extremely effective mechanism in the case of activities for which complete information exists and future uncertainty is manageable. In some circumstances it is possible to incorporate some mechanisms into the contract to create flexibility and allow for future uncertainty. The rationale for building flexibility into an outsourcing contract is based on the premise that factors both internally and externally may change and thus impact the achievement of the desired objectives of the outsourcing. For example, the internal requirements of the sourcing organisation may change during the outsourcing contract or another supplier in the supply market may achieve a technology breakthrough, which allows it to realise significant performance improvements. In the latter case, the establishment of a long-term contract with a competing supplier prevents the sourcing organisation from accessing the superior capabilities of this supplier. Therefore, incorporating elements into a contract that create flexibility can ensure that the desired benefits are being achieved from outsourcing and in particular, ensure that the sourcing organisation is not locked into a relationship with an uncompetitive supplier. Flexibility can be achieved through either incomplete or incentive contracts (Harris *et al.*, 1998). Incomplete contracting creates a situation in which parts of the contract can be renegotiated based upon changes in circumstances. There are a number of methods of incorporating flexibility into a contract through incomplete contracting.

- *Price flexibility* – allows prices to be renegotiated as circumstances change during the contract. Incorporating price flexibility means that all future contingencies do not have to be fully considered at the outset, as the buyer and supplier are aware that prices can be adjusted to reflect changes in circumstances. For example, changes in the requirements of the sourcing organisation during the contract may necessitate an adjustment in prices.
- *Renegotiation* – mechanisms are incorporated into the contract that allow for renegotiation based upon changes in the business environment. The contract may include specific clauses under which renegotiation should occur including fixed calendar dates or changes in economic indices. Renegotiation often involves renegotiating more than price and can also focus on the terms of contract.
- *Contract length* – the employment of shorter contracts can be employed to achieve flexibility. At the end of the contract period a new contract can be negotiated that reflects the current circumstances both internally and externally.
- *Early termination* – a clause may be incorporated into the contract that sets out the conditions under which the contract may be terminated. The omission of such a clause can result in considerable penalties in the event of the contract being terminated prematurely.

Incentive contracting involves incorporating mechanisms into the contract that allows the supplier to share any cost savings or profits generated through the outsourcing relationship. Incentive contracts are often employed to encourage performance improvements in the outsourcing arrangement in areas such as cost reduction and service levels. The contract will include mechanisms that ensure the supplier shares any savings that are realised from performance improvements. Alternatively, the supplier may be penalised through reduced payments in situations where the required performance levels are not met. However, in many cases it is not possible to design a contract sufficiently flexible to account for all future contingencies. For example, in the outsourcing of a very straightforward activity with stable demand requirements, a contract can be a very effective means of driving the relationship. However, in many cases there is a strong argument for developing relationships that foster collaboration to compensate for any gaps in the contract. In the discussion on the role of the SLA, it has been argued that in the case of a collaborative relationship, rigid adherence to the contract can be detrimental to the development of the relationship. In such circumstances, it is more beneficial to establish relational mechanisms such as a joint problem-solving culture to improve any deficiencies in service levels from the supplier. For example, in the management of quality it is much better for the buyer to assist the supplier in improvement efforts rather than impose penalties on the supplier as set out in the contract. Such an approach allows the relationship to develop over the long term as the individuals at the buyer–supplier interface build up an understanding of each other's requirements. However, such an approach emphasises the importance of selecting a supplier that possesses the capabilities to meet the performance levels required by the buyer. In fact, the performance of the supplier and nature of the relationship will be the key determinants influencing how much emphasis is placed on the contract. Although the contract will play a role, the health of the relationship will depend primarily upon how it is allowed to develop through the employment of relational mechanisms to promote greater understanding at the buyer–supplier interface. Therefore, in the case of a collaborative relationship, any potential problems that may arise, not allowed for in the contract can be addressed through the employment of relational mechanisms.

Illustration 10.2

Formal contracts and relational contracting as complements
Laura Poppo and Todd Zenger carried out a study testing whether relational contracting and formal contracts would function as complements or substitutes using data collected from outsourcing relationships in the area of information systems services. Their study involved surveying senior managers in relation to the sourcing of information services including data entry, software application development, network design and maintenance. The prevailing view in much of the literature is that relational contracting acts as a substitute for formal and explicit contracts, the assumption being that relational mechanisms, which include trust and co-operation are more effective and less costly alternatives

to formal contracts. Furthermore, the presence of formal contracts between the parties in the relationship is likely to undermine trust and encourage opportunism. However, the authors in this study challenge this view and propose that formal contracts and relational contracting can operate as complements. They argue that clear and precise contracts can actually encourage higher levels of co-operation, a long-term perspective, and greater trust in the relationship. For example, the presence of a well-specified contract promotes a longer-term perspective through increasing the penalties that are incurred from dissolving the relationship. Clear contractual safeguards promote the expectations that each party to the relationship will function on a co-operative basis and therefore complements the informal limitations of relational contracting. They also argue that the long-term perspective and co-operation encouraged through relational contracting is likely to compensate for any unexpected difficulties or complications not accounted for in the formal contract.

The findings in the study confirmed much of the authors' initial propositions that formal contracts and relational contracting function as complements. In particular, they found that companies tend to employ more relational mechanisms as contracts become more customised and they employ more complex contracts as they develop more collaborative relationships. Customised contracts include contingencies, adaptive processes, and controls that limit the potential for opportunistic behaviour from each party in the relationship. Their findings also suggest that complex contracts and relational contracting serve as complements when explaining each party's satisfaction with the performance of the relationship. Satisfaction with relationship performance was measured along the dimensions of the overall cost of the service, the quality of the service and the responsiveness of the supplier to problems or queries. Contract complexity indirectly enhances relationship performance by enhancing relational contracting, which in turn increase relationship performance. Higher levels of relational contracting also positively impact contract complexity, which in turn enhances relationship performance.

Their findings in relation to the influence of technology on formal contracts and relational contracting are also interesting. Although companies did not employ relational contracting as a result of increasing levels of specialised assets in the exchange, they found that the presence of technological change and specialised assets led to the employment of higher levels of relational contracting. In particular, there was found to be a high correlation between the presence of technological change and relational contracting. The companies studied appeared to use relational mechanisms in response to the rapid changes associated with information services technology. Customised contracts were also less likely to be used in the case of information services that were characterised by difficult performance measurement and technological change. The authors argue that the overall findings confirm the complementary nature of formal contracts and relational contracting because both have unique origins. Where both of these governance modes to have similar origins, they would function as substitutes. However, there are a number of important qualifications that are made in relation to the generalisability of the findings from this research. The institutions of nations and legal systems will influence the effectiveness of the employment of formal contracts as governance modes. Therefore, these findings cannot be generalised to countries with a lack of cultural and legal commitment to formal contracts. Moreover, to strengthen the findings of this study the relationships would have to be analysed over time to more fully test the dynamics of the complementary nature of formal contracts and relational contracting. The authors stress that many of the outsourcing relationships studied were in the early stages of development and that it was unlikely that the informal mechanisms such as trust and cooperation associated with relational contracting had developed. The logic of this argument is that formal contracts will become less important over time as trust and cooperation are allowed to develop. Therefore, formal contracts may be important in the initial stages of the relationship, but diminish in importance as more relational mechanisms develop.

Source: Poppo, L. and Zenger, T. (2002). Do formal contracts and relational governance function as substitutes or complements? *Strategic Management Journal*, **23**, 707–25.

10.4   Managing the relationship

10.4.1   Managing the interaction process

The direction and management of the buyer–supplier relationship will be influenced by the objectives of the outsourcing strategy. The importance of the activity and the level of risk in the supply market will also have determined the type of relationship adopted. However, the success of the relationship will be very much determined by how it is managed at the operational level. Therefore, it is critical that both the buyer and supplier have the necessary skills and resource to manage the interaction process at the operational level. The interaction process will differ considerably depending upon the type of relationship adopted. For example, managing a close collaborative relationship requires substantial investments from both the buyer and supplier. Such an interaction process differs considerably from that of the adversarial relationship as shown in Figure 10.1. Under the adversarial relationship the interface is confined to salesperson and buyer with limited involvement from the strategic level of each

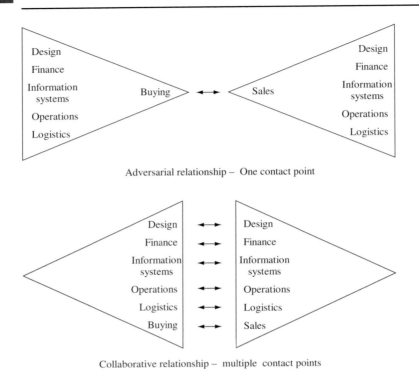

Figure 10.1 Interaction patterns in adversarial and collaborative relationships

business. This interface structure can be established with a limited commitment of resources and can be dissolved relatively easily. However, under a close collaborative the relationship is multi-functional with bi-directional information sharing and problem solving. Direct interactions between a range of business functions reinforce the philosophy of collaboration that has to occur in the relationship.

The interaction patterns between the relevant participants at the buyer–supplier interface are critical in ensuring the development of a close collaborative relationship. This will involve a significant resource commitment and will also rely on input from senior management at various stages in the development of the relationship. In the early stages of the relationship interaction is important to enable a good rapport to develop between the people responsible for managing the relationship. Meetings between the top management of both the buyer and supplier will take place to demonstrate the high level of commitment to the relationship. Visits by the personnel of both the buyer and supplier to their respective sites is essential in creating a mutual understanding of each party's needs and capabilities. Face-to-face interaction is most successful means of understanding the needs of both parties.

Illustration 10.3

Japanese supply chain superiority

A study carried out by Jeffrey Liker and Yen-Chun Wu provides an illustration of how some companies can realise better performance levels from suppliers than their competitors. Their study focused on suppliers that supplied both Japanese car makers and the Big Three car makers in the US. In the 1980s and 1990s, many of the Japanese car makers such as Honda, Nissan and Toyota established manufacturing plants in the US and quickly disseminated their world-class manufacturing standards to many of the local US suppliers. As well as implementing just-in-time type delivery systems, they also encouraged the use of lean manufacturing techniques. Lean manufacturing involves delivering the highest quality product at the lowest cost on time. This involves analysing all the steps in the process of converting raw materials into an end product for the customer and eliminating any steps that create waste. The objective is to minimise waste so that the product can flow as efficiently as possible. By adopting this philosophy with suppliers in the US, the Japanese car makers were able to obtain performance levels comparable with those of their suppliers in Japan. The performance of these suppliers illustrates clearly the positive impacts that the customer can have upon these suppliers. The study found that local US suppliers performed at a much higher level when they were supplying Japanese car makers than when they were supplying the US car makers. In particular, the study found that a number of policies pursued by the Japanese car makers have been central to the achievement of these higher performance levels.

- The Japanese transplants have more level production schedules in order to avoid rapid changes in demand, thus allowing suppliers to hold less inventory. This is clearly illustrated in the example of Johnson Controls, a seat manufacturer. Prior to working with Toyota it would hold a large inventory of seats. After working closely with Toyota in a lean manufacturing initiative, inventory levels dropped from 32 days to 4.1 days. This involved a number of changes in Johnson Control's manufacturing practices including the reducing of die setup times from 6 hours to 17 minutes.
- The Japanese transplants have created a disciplined system of delivery time windows during which all parts must be received at the assembly plant. They also provide more predictable delivery schedules. Their order demand is more stable reflecting leveller schedules in their assembly plants. Other customers with non-level production schedules create problems in a number of areas for their suppliers including higher inventory costs and higher transportation costs associated with expedited deliveries.

- Japanese transplants have developed lean transportation systems to deal with mixed-load and smaller delivery lot-sizes. They have developed closer relationships with transportation carriers, set stringent delivery requirements, and have compensated for longer distances in the US by adopting innovative delivery methods and efficient loading methods. By focusing on a small number of carriers, only one in the case of Toyota, they have also obtained more reliable service through tightly scheduled deliveries, shipment tracing and effective communication. This contrasts with the practices of some of the US car makers. For example, GM uses seven carriers. The study found that GM experienced a much higher level of delayed deliveries and lower on-time pick-ups than Toyota. The use of so many transportation carriers has made it difficult for GM to integrate them into its own scheduling systems and minimise late deliveries.
- Japanese transplants encourage suppliers to ship only what is needed by the assembly plant at a particular time, even if this means partially filled trucks. This contrasts with Ford who place pressure on suppliers to fully load their trucks. Ford believes that by using all the space in its trucks it will lower its transportation costs. However, such a policy encourages suppliers to fill trucks whether or not it actually needs the parts and creates additional costs in the area of inventory.

It must be stressed that the US companies were not found to be weak along all the measures analysed. In fact, the analysis found that they were achieving similar performance levels in certain areas in relation to the Japanese companies. This was in part as a result of the US car makers beginning to adopt some of the practices pursued by the Japanese transplants.

Source: Liker, J. K. and Wu, Y. C. (2000). Japanese automakers, US suppliers and supply chain superiority. *Sloan Management Review*, **41**, No.1, 81–93.

10.4.2   Assigning a relationship promoter

In the case of a close collaborative relationship a relationship promoter can be appointed to break down any potential barriers to the development of the relationship. A relationship promoter can either be a single person or a team of people. The relationship promoter must identify evidence of a supportive, trusting and committed relationship between the buyer and supplier. Gemunden and Walter (1994) have identified the following barriers to collaboration that the relationship promoter should address.

- *'No knowledge of each other'* – each party in the relationship may have limited awareness of the roles of other parties in the relationship. For example, a buyer-centred view of the relationship will lead to an ignorance of the problems and challenges faced by the supply-side of the relationship. Therefore, the role of the relationship promoter will involve informing the participants at the buyer–supplier interface of the holistic nature of the relationship.

- *'No ability to co-operate'* – each party does not understand how to co-operate with their counterparts. The relationship promoter must establish a means of enabling the parties to achieve a consensus and a common viewpoint.
- *'No will to co-operate'* – this is the most difficult barrier to overcome. The relationship promoter must attempt to work towards a change in the people's attitudes and to support them in overcoming emotional resentments to deal rationally with the situation.

In fact, the development of collaborative relationships constitutes major changes for organisations. Close collaboration implies a radical change in the way people work in contrast to the adversarial model; for example there is increased emphasis upon team working, joint decision making and collaborative activity between buyer and supplier organisations. The more radical these changes are, the more difficult the task of implementing them is likely to be. It implies changes to the social systems of at least two separate organisations with the scope for resistance being considerable (Boddy *et al.*, 1998). McIvor and McHugh (2000) have highlighted a number of organisation change implications for organisations attempting to develop collaborative relationships with suppliers.

- The changes effected by collaborative arrangements strike at the very heart of the organisation and have implications for the way in which an organisation is structured, individual roles, responsibilities, reward systems and reporting relationships. The changes are systemic in that modification to structural arrangements for example, automatically has a knock-on effect upon individual roles, responsibilities, reward systems and reporting relationships. It is essential to ensure the adoption of a holistic approach to managing the entire process. This is likely to demonstrate acknowledgement of the integration that exists between the different components of the organisational system.
- The pursuit of collaborative buyer–supplier relations requires a major shift in the mindset or operational paradigm of organisational members. In other words, it requires cultural change within the buyer and supplier organisation where there is enhanced understanding of the concept of collaboration. It is acknowledged that effecting culture change is often regarded as being a mammoth task, which is made even more difficult by the deeply embedded culture that has evolved over a long period of time.
- A company will have considerable difficulties adopting collaborative arrangements with external suppliers if it cannot develop a collaborative mindset across internal functions. A culture in both the customer and supplier organisation must exist in order to facilitate and encourage joint problem solving and decision-making across intra-organisational boundaries. It is not enough to

change the attitudes of the purchasing personnel but the attitudes of the other business functions and senior management must also be changed in the pursuit of collaborative buyer–supplier relations. It requires a culture permeating the organisation hierarchy that encourages and values collaboration. However, such a requirement presents organisations with a challenge due to the embedded culture of both the buyer and supplier, particularly if they have traditionally operated on an adversarial basis.

• Senior management as organisational leaders have a critical role to play in bringing about the required transformation in culture required for collaborative buyer–supplier relationships. Senior management must have an awareness and knowledge of working processes, practices and relationships throughout the organisation. This will create an understanding of the implications of their decisions in relation to the support structures necessary to facilitate implementation. Thus, considering the inter- and intra-organisational implications of collaboration, to effect successful collaborative relationships, it is essential that organisations pursue a much more participative approach to the strategy-making process.

• When adopting more collaborative buyer–supplier relationships it is essential to assess the impact that this will have upon those who will be most affected by the impending change. Effective collaboration between individuals and groups places a newfound emphasis upon skills that were not so necessary when engaging in the more traditional adversarial relationship. In particular, assessing the impact on those most affected by new developments is likely to reveal a need for training to permit skill acquisition and development in a variety of areas including team working, problem solving, negotiation and conflict management.

• To achieve the benefits associated with buyer-supplier collaboration it is essential that all of those involved are committed to the strategy. This must be supported by an organisational programme that aims to enable individuals acquire and develop skills, behaviours and attitudes to facilitate the implementation of strategy. As a pre-requisite to programme development and delivery, it is essential that each of both the buyer and supplier identify the range of skills, behaviours and attitudes that are critical to support effective collaborative activities.

10.5   Relationship performance evaluation

The importance of relationship management in the context of outsourcing has already been emphasised. Much of the risk associated with outsourcing arises from the failure of suppliers to deliver and meet the requirements of the buyer. In order to reduce these risks and pre-empt supplier failure, the buying organisation must have a formal mechanism to determine whether the supplier is meeting the performance levels set and whether the objectives in its approach to relationship management are being achieved. There are a number of aspects that should be considered in this evaluation including supplier performance, the strength of the relationship and the level of dependency.

10.5.1 Supplier performance

The supplier selection process is concerned with determining whether the supplier has the capability to deliver to the required standards. Evaluation of supplier performance is concerned with determining whether the supplier is delivering to the required standards during the contract. This analysis will focus on performance metrics related to quality, delivery, service and ability to reduce costs. Kannan and Tan (2002) have provided some criteria that can assist in performance evaluation as shown in Table 10.2. Having an effective mechanism of evaluating supplier performance can also serve as basis for comparing performance levels with that of other potential suppliers in the supply market. The approach to evaluation will depend upon the nature of the relationship. In the case of an adversarial relationship, the buyer will focus on measuring the performance of the supplier quantitatively along of number of criteria including price, quality, and delivery. As with the supplier selection decision in an adversarial relationship, the dominant criteria in the evaluation will be price and cost reduction. The focus is on inspecting the outcomes of the process rather than attempt to diagnose the causes of poor performance (MacBeth, 1994). The responsibility for performance rests solely with the supplier with little or no assistance provided from the buyer to resolve problems.

Price is only one element of the costs associated with sourcing an activity from a supplier. In the case of critical activities, the total costs associated with sourcing the activity from a particular supplier should be considered. A useful technique that can support this process of evaluation is total cost of ownership. Total cost of ownership (TCO) refers to all the costs associated with the acquisition, use, and maintenance of a product or service (Ellram, 1993). TCO considers costs all the way from idea inception, as in working with a supplier to develop a new or improved product or service, through to warranty claims associated with that product or service once it is in use by the sourcing organisation. Therefore, rather than focusing on price alone, TCO attempts to determine all the cost elements of sourcing in order to identify opportunities for cost reduction and the elimination of inefficiencies. TCO analysis focuses on the interfaces between the sourcing organisation and the supplier. As shown in Figure 10.2, the TCO approach proposed by Ellram (1993) segments the analysis of costs into the following three components.

Table 10.2 *Supplier performance evaluation criteria (Kannan and Tan, 2002, Table IV, p. 16)*

Dimension	Criteria
Strategic commitment of supplier to the buyer	• Willingness to integrate supply chain management relationship • Supplier's order entry and invoicing system, including electronic data interchange • Supplier has strategic importance to buyer firm • Supplier's effort in promoting JIT principles • Buyer annual orders as a percentage of overall business • Supplier's ability to make a decent profit for supplying buyer • Supplier's willingness to share confidential information
Ability to meet buyer needs	• Ability to meet delivery due dates • Honest and frequent communications • Commitment to quality • Commitment to continuous improvement in product and process • Reserve capacity or the ability to respond to unexpected demand • Flexible contract terms and conditions • Financial stability and staying power
Capability	• Technical expertise • Industry knowledge • Scope of resources • Testing capability
Buyer–supplier fit	• Geographical compatibility/proximity • Cultural match between the companies • Past and current relationship with supplier
Honesty and integrity	• Insurance and litigation history • Open to site evaluation • Supplier's effort in eliminating waste

From the Institute for Supply Management[TM]. Reproduced with permission.

- *Pre-transaction costs* – the costs that are incurred prior to receiving the product or service, and even prior to placing the order. They include all the costs incurred from the time that the sourcing organisation considers the possibility of sourcing the product or service, up to, but not including, order placement. For example, potential costs in this category include those of investigating potential suppliers, qualifying and educating suppliers in relation to the systems and expectations of the sourcing organisation, and adapting to the systems and delivery methods of the supplier.

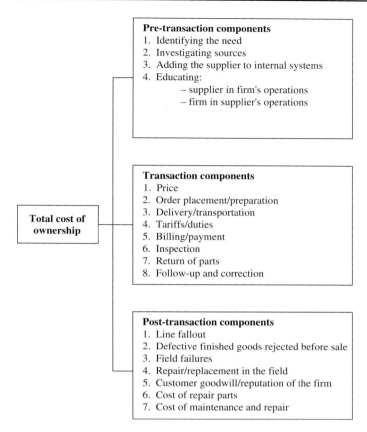

Figure 10.2 Major categories for the components of total cost of ownership in a manufacturing context (Ellram, 1993, Fig. 1, p.7) From the Institute for Supply Management™. Reproduced with permission.

- *Transaction costs* – are related to order placement and receipt, and include the price of the product or service. These transaction costs include the costs associated with preparing and placing the order, following up on the order, receiving, matching receiving data to the invoice, and paying the bill. These cost elements are more widely recognised than pre-transaction and post-transaction costs, because they are closely associated with the transaction itself. Potential omissions in transaction costs include purchase order preparation, order matching and receiving and invoice payment. However, the implementation of IT systems can reduce the costs of many of these elements.
- *Post-transaction costs* – occur after the product or service is owned or in use by the sourcing organisation. These costs may arise after the order is received, or in the future when the product or service is in use or being modified, repaired, or disposed of. The further in time a cost occurs from the transaction, the less likely the cost will be related to the sourcing of a product or service from that

particular supplier. Post-transaction costs often overlooked include product repair in the field, routine and special maintenance costs, costs associated with replacement part scarcity or obsolescence.

While it is not possible for the sourcing organisation to identify precisely all the costs associated with each of the elements within these categories, the approach provides a useful framework for performance analysis and guidance. It can assist in analysing the costs of doing business with a particular supplier over the lifetime of the relationship. It is not only a framework for analysing costs on the supplier side but can assist in identifying areas in the sourcing organisation where costs and inefficiencies can be eliminated in order to build a stronger relationship with the supplier. It is also possible to identify how specific requirements of the sourcing organisation are increasing the costs of suppliers.

In the case of a collaborative relationship, the approach to evaluation differs considerably. The importance of measuring and monitoring performance in a collaborative relationship is emphasised by the increased dependency between the buyer and supplier. The evaluation of performance is a joint process with both the buyer and supplier attempting to identify and deal with the causes of poor performance in any areas. With both the buyer and supplier being responsible for the success of the relationship, the focus is on improvement. This focus on improvement not only centres on cost reduction but also can encompass any area of operations. There is also an onus on the buyer to achieve improvements. The buyer may attempt to achieve improvements that can assist the supplier in meeting the required performance levels. Therefore, both the buyer and supplier must be aware of each other's expectations of performance. The development of close collaborative relationships will involve the buyer and supplier working jointly over the long-term to improve performance levels and meet each other's expectations.

10.5.2   The strength of the relationship

As well as evaluating supplier performance, the nature of the relationship with the supplier must be evaluated. Evaluation of the strength of the supply relationship will be guided by the initial objectives established for the outsourcing process. This evaluation is most appropriate in the context of collaborative relationships. For example, monitoring the strength of a close collaborative relationship will involve analysing the presence of intangible factors such as joint problem solving, high levels of information exchange and top management commitment from both the buyer and supplier. Table 10.3 provides a useful categorisation of the factors that describe the strength of the relationship. The strength of the relationship describes the factors that create bonds between the buyer and supplier. The *economic factors* describe the level of dependency of the supplier on the buyer. Exit costs relate to

Table 10.3 *Factors indicating the strength of the relationship (adapted from Olsen and Ellram (1997) and Kannan and Tan (2002))*

Factor
Economic factors
(1) Volume or financial value of purchases as an overall percentage of supplier turnover
(2) Strategic importance of the buyer to the supplier
(3) Exit costs
Character of the exchange relationship
(1) Types of information exchanged
(2) Willingness to share information
(3) Frequency of communication
(4) Level and number of personal contacts
(5) Duration of the relationship
Co-operation between buyer and supplier
(1) Co-operation in areas such as new product development and cost reduction
(2) Willingness to integrate systems
(3) Integration of management
Distance between the buyer and supplier
(1) Social distance
(2) Cultural distance
(3) Technological distance
(4) Time distance
(5) Geographical distance

the level of investments made by the supplier that are specific to the needs of the buyer. These high exit costs limit the ability of the supplier to transfer these investments to other buyers. The *character of the exchange relationship* describes the characteristics that create stronger bonds between the buyer and supplier. For example, a high frequency and wide range of information exchange between the participants at the buyer–supplier interface can facilitate the development of a strong relationship. Moreover, these bonds will be further strengthened if these exchange mechanisms have been developed over a long period of time. *Co-operation between the buyer and supplier* indicates the level of collaboration at various levels at the buyer–supplier interface. This can also include co-operation in area of integrating systems to facilitate greater transparency between the buyer and supplier. *Distance between the buyer and the supplier* is composed of five factors (Ford, 1984). Social distance relates to the familiarity of the individuals who manage the relationship in both the buyer and supplier. Cultural distance refers to the differences between the norms and values of the buyer and supplier.

Technological distance relates to the differences between product and process technologies of the buyer and supplier. Time distance describes the cycle time between order placement and order fulfilment of the product or service involved.

Finally, geographical distance refers to the physical distance between the buyer and supplier.

10.5.3 The level of dependency

In the choice of relationship, the level of dependency was influenced by the importance of the activity being outsourced and the number of suppliers in the supply market. For example, in the case of a peripheral activity with a readily available number of suppliers, the most appropriate strategy was to create a low level of dependency on the supplier through employing an adversarial relationship. However, changes in the internal and external environment can have a positive or negative effect upon the level of dependency for the buyer over the lifetime of the relationship. There are a number of factors that can influence the level of dependency.

- *Supply market* – changes in factors such as the number of available suppliers, competitive demand for supply and the performance of other suppliers can impact the level of dependency in the relationship. For example, when establishing the relationship with the supplier, the sourcing organisation may have established a close collaborative relationship because there were only a limited number of competent suppliers in the supply market. However, over time changes such as lower entry barriers and advances in technology can lead to more competent suppliers entering the supply market. In this case, the sourcing organisation will have to consider switching to another more competent supplier or altering the level of dependency with the existing supplier by reallocating some of the business to another supplier.
- *Activity importance* – the importance of the activity provides a reliable indication of the level of resource and attention that should be given to managing the supply relationship. Over time the importance of the outsourced activity can increase or diminish. For example, in a manufacturing context, advances in technology can erode the importance of certain components. Developments in the supply market such as the entry of more competent suppliers or acquisition activity by suppliers can also affect the importance of the activity. Again, the sourcing organisation will have to consider altering the level of dependency in the relationship to reflect changes in the importance of the activity. Alter natively, the level of dependency can be altered by reducing the importance of the activity. For example, the sourcing organisation may decide to redesign a product to reduce the need for a critical component that is prone to failure in the supply market.

The evaluation of these aspects of relationship performance will be linked to the objectives of the outsourcing strategy. Evaluation of relationship performance should serve as a stimulus for action. The following are potential actions that can result from this evaluation.

- *Maintain the relationship at its current level* – in this case, the buyer maintains the relationship while not allocating additional business to the supplier.
- *Further develop the relationship* – in this case, the outsourcing of the activity has proved to be successful with the supplier having met and exceeded the performance objectives. As a result, the buyer may further strengthen the relationship through giving the supplier more business. Relational mechanisms such as the further integration of information systems and greater collaboration in the design process can also be employed to further develop the relationship. In the case of an adversarial relationship in which the supplier is very competitive, the relationship can be developed through giving the supplier more business. The buyer may also change the nature of the relationship it has adopted with the supplier. For example, the buyer may decide to establish a collaborative relationship with a supplier with which it has had an adversarial relationship in the past because the supplier has performed to a very high standard and shown a willingness to develop a collaborative relationship. However, the effects of further developing the relationship on the level of dependency must be carefully considered.
- *Discontinue or reduce the scope of the relationship* – the buyer may decide to discontinue the relationship with the supplier. If possible, the business should be switched to one or a number of other suppliers. As a last resort, the buyer may decide to bring the business back in-house – sometimes referred to as back-sourcing. Clearly, the length of the contract with the supplier will be a key determinant on the ease with which an activity can be switched to another supplier or brought back in-house. For example, short-term contracts can be employed for certain outsourced activities that allow the relationship to be discontinued if the supplier fails to meet the required levels of performance. However, for more complex business activities outsourcing decisions can be extremely difficult and costly to reverse. Indeed, many organisations do not put contingency plans in place to allow for outsourcing failure. Barthelemy (2003) has found that organisations that do not plan for outsourcing failure often overlook factors including material reversibility clauses (i.e. the option to buy back equipment from the supplier) – or human reversibility clauses (i.e. the option to re-hire employees transferred to the supplier) in the contract. Failure to consider these issues makes it extremely difficult to discontinue the outsourcing agreement and in particular reduces the power of the sourcing organisation in any negotiations with the supplier. However, any decision to discontinue the relationship should not be taken lightly and only after significant attempts to reconcile any problems affecting the relationship. Moreover, the buyer may decide to gradually reduce the scope of the relationship with the supplier and transfer some of the business to another supplier over a period of time.

Illustration 10.4

Offshore supply management

Establishing outsourcing arrangements with offshore product and service providers can often appear on the surface to be a very attractive proposition for organisations. Organisations can access highly skilled labour at a fraction of the cost available locally in a range of areas including contract manufacturing, advertising services, transaction processing, insurance and legal services. There are a number of ways in which organisations can manage these arrangements. An organisation may establish and manage a subsidiary in a foreign location in order to avail of local skills and lower labour costs and also establish a presence in the offshore market. For example, companies have established research and development subsidiaries in India and China in order to reduce costs and speed up development times. Another potential arrangement involves sourcing products or services from independent suppliers on an adversarial or collaborative basis. Furthermore, an organisation may request its local suppliers to build, operate and transfer facilities to a foreign location. A further adaptation of this arrangement involves the customer and supplier establishing a joint venture, which allows the supplier to market their products or services to other customers. As in the case of outsourcing locally, the types of activities and motives for outsourcing will influence the sourcing arrangements adopted. However, the decision in relation to offshoring is further complicated as a result of the additional risks associated with managing suppliers in an international context including:

- exchange rate volatilities can limit any anticipated cost savings;
- cultural and language differences may hinder the ability of offshore suppliers to deliver the service levels required;
- geographical distance still poses a significant challenge to co-ordinating and managing business activities;
- non-tariff barriers can be employed to avoid international trade agreements by placing restrictions on companies selling outsourced services locally;
- government instability can threaten outsourcing arrangements; and
- countries with underdeveloped legal systems can prevent the changing of outsourcing contracts.

Although some of these risks are uncontrollable, organisations can adopt a variety of sourcing configurations with offshore suppliers in order to alleviate the potential negative impacts of these risks. For example, when offshoring peripheral activities organisations can often attain cost savings as a result of favourable exchange rates. In this case, a flexible sourcing arrangement should be adopted in which an organisation can switch supply to other offshore locations or local sources in order to limit the effects of fluctuations in exchange rates. In relation to the offshoring of more critical activities, obtaining cost reductions may be only one of a number of motivations that also includes service improvements and reduced time-to-market for its products. In this case, a more collaborative and longer-term relationship will be adopted, where for example, fluctuations in exchange rates may be considered as a temporary problem that can be resolved to the satisfaction of both the buyer and supplier rather than leading to the termination of the relationship. Furthermore, companies may establish offshore manufacturing or service operations in order to circumvent protectionist measures and establish a foothold in offshore markets. The approaches to global sourcing adopted by US and Japanese firms provide an illustration of these issues. Throughout the 1980s and 1990s, US companies were motivated primarily by cost considerations through global outsourcing, whilst Japanese companies employed global sourcing as a means of enhancing quality, reliability and accelerating new product development. This is reflected in the predominance of US companies sourcing from independent foreign suppliers whilst Japanese companies source mainly from offshore subsidiaries or affiliates. For example, in South-east Asia Japanese companies built transplants for manufacturing components and final products and transferred elements of the *keiretsu* into relationships with suppliers. The Japanese companies shared knowledge through engineers and designers transferring on a regular basis from their home plants to offshore plants. As well as accessing lower resource costs for these transplants, Japanese companies exported their products from these South-east Asian countries to the US in order to avoid some of the US protectionist measures they encountered when exporting directly from Japan.

Although companies can adopt a range of sourcing arrangements in order to leverage the knowledge and capabilities available in offshore locations, there are still considerable risks. Many organisations have pursued offshoring in a piecemeal fashion in the pursuit of short-term cost savings or quality improvements. As with domestic outsourcing focusing on cost savings via for example, favourable exchange rates or lower labour costs, as the dominant motive for offshoring is not a strategy for achieving competitive advantage. Furthermore, the popular perception of offshoring is that it involves only low-value and highly labour intensive activities. As offshore locations such as India and China develop their capabilities in a range of business activities, companies have the opportunity to leverage these capabilities in order to enhance their own product and service portfolio. However, in order to avail of these potential benefits offshoring should be placed in its proper strategic context, which allows an organisation to exploit its own competitive strengths and leverage the comparative advantages that exist in a range of global locations. Clearly, these comparative advantages can be in the form of both lower resource costs and the leveraging of knowledge and capabilities that exist in offshore locations. In order to achieve these potential benefits an organisation should adopt a range of relationship strategies – ranging from wholly owned subsidiaries to adversarial arrangements with independent suppliers – which are influenced by the importance of the activity and the number of available suppliers.

Sources: Kotabe, M. and Murray, J. Y. (2004). Global sourcing strategy and sustainable competitive advantage. *Industrial Marketing Management*, **33**, 7–14.
Schneiderjans, M. J. and Zuckweiler, K. M. (2004). A quantitative approach to the outsourcing–insourcing decision in an international context. *Management Decision*, **42**, No. 8, 974–986.

REFERENCES

Barthelemy, J. (2003). The seven deadly sins of outsourcing. *The Academy of Management Executive*, **17**, No. 2, 87–98.

Blumberg, D. F. (1998). Strategic assessment of outsourcing and downsizing in the service market. *Managing Service Quality*, **8**, No. 1, 5–18.

Boddy, D., Cahill, C., Charles, M., Fraser-Jraus, H. and MacBeth, D. (1998). Success and failure in implementing supply chain partnering: an empirical study. *European Journal of Purchasing and Supply Management*, **2**, No. 2/3, 143–51.

DiRomualdo, A. and Gurbaxani, V. (1998). Strategic intent for outsourcing, *Sloan Management Review*, **39**, No. 4, 67–80.

Ellram, L. (1993). Total cost of ownership: elements and implementation. *International Journal of Purchasing and Materials Management*, **29**, No. 3, 3–11.

Ellram, L. M. (1990). The supplier selection decision in strategic partnerships. *International Journal of Purchasing and Materials Management*, **26**, No. 3, 8–14.

Ford, D. (1984). Buyer/seller relationships in international industrial markets. *Industrial Marketing Management*, **13**, 101–12.

Gemunden, H. and Walter, A. (1994). The relationship promoter – key role for inter-organisation processes. In: *Relationship Marketing: Theory Methods, and Applications,* Sheth, J. and Partratiyer, A. (eds.). Atlanta, GA: Emory University, Centre for Relationship Marketing.

Harris, A., Giunipero, L. C. and Hult, G. (1998). Impact of organisational and contract flexibility on outsourcing contracts. *Industrial Marketing Management*, **27**, 373–384.

Hiles, A. N. (1994). Service level agreements: panacea or pain? *The TQM Magazine*, **6**, No.2, 14–16.

Kannan, V. R. and Tan, K. C. (2002). Supplier selection and assessment: their impact upon business performance. *The Journal of Supply Chain Management*, **38**, No. 3, 11–21.

Larson, K. D. (1998). The role of service level agreements in IT service delivery. *Information Management and Security*, **6**, No.3, 128–32.

Lee, M. (1996). IT outsourcing contracts: practical issues for management. *Industrial Management and Data Systems*, **1**, 15–20.

McIvor, R. T. and McHugh M. L., (2000). Partnership sourcing: an organisation change management perspective. *The Journal of Supply Chain Management*, **36**, No. 3, 12–20.

Olsen, R. F. and Ellram, L. M (1997). A portfolio approach to supplier relationships. *Industrial Marketing Management*, **26**, 101–13.

MacBeth, D. K. (1994). The role of purchasing in a partnering relationship. *The European Journal of Purchasing and Supply Management*, **1**, No. 1, 19–25.

11 Case study – outsourcing experiences at Telco

11.1 Introduction

This case study focuses on a telecommunications equipment manufacturer. The telecommunications industry has been characterised by extensive outsourcing by original equipment manufacturers (OEMs) over the last few years due to issues such as shortening product cycles and time to market pressures. Many OEMs in this industry have increasingly become 'systems integrators'; namely specialists in the management and co-ordination of production and service providers. This case study identifies the drivers that have influenced outsourcing in the telecommunications industry. The case organisation has outsourced many activities to suppliers including assembly operations, manufacturing, logistics and design. A central part of this strategy has been the adoption of more collaborative relationships with its key suppliers in order to reduce the risks associated with outsourcing. The company will be referred to as Telco for purposes of confidentiality. The implications of the outsourcing strategy are drawn from an analysis of this company and three cases of outsourcing evaluation and management over an 18-month period. These three case studies are related to the stages in the outsourcing framework presented in this book in order to illustrate both its explanatory and prescriptive nature. A critical evaluation of Telco's experience with outsourcing is carried out and in particular, its attempts to develop collaborative buyer–supplier relationships. The implications of the case organisation adopting a modular approach to the arrangement of its supply base are examined. Some of the issues developed in this case study are for illustrative purposes only.

11.2 An overview of Telco

The case organisation had experienced considerable change over recent years. It had been taken over by a multi-national telecommunications company resulting in an increasing emphasis on the adoption of world-class practices such as a total quality culture, integrated product development and continuous improvement.

Table 11.1 *The de-layering of the organisation*

Layer	1985	Currently
Layer 1	Director	Managing Director
Layer 2	Senior managers	Senior managers
Layer 3	Managers	Managers
Layer 4	Superintendents	Team leaders
Layer 5	Fore-persons	Team members
Layer 6	Supervisors	
Layer 7	Charge hands	
Layer 8	Workforce	

The take-over had resulted in major changes in the management structure of the organisation. The organisation flattened its management structure, with all employees encouraged to take responsibility for improvement in their own work areas. The individual senior management team members have functional and critical process responsibilities that they carry out through managers and team leaders. The organisation has three less layers than it had in 1985, as shown in Table 11.1. The organisation has a team-based structure in which all employees are encouraged by their team leaders and managers to become involved in resolving problems and improving quality in their own work areas. Personal development training courses have been introduced since the introduction of the total quality ethos to the organisation. These are used to promote continuous improvement methodologies and to encourage involvement in teams.

Telco designs and manufactures a range of transmission and switching products and provides a total solution for its customers. In most cases the product is contained in a telephone exchange type rack that houses the sub-racks for each multiplexor. The multiplexor is the heart of the product providing the capability to form the network that carries the communications traffic to the customer's specification. A product may contain a number of multiplexors. Multiplexors take data being transmitted at a specific speed and interleave it into a faster bit stream for long-distance and high-speed transmission. The multiplexors are at the heart of the systems that provide the backbone of telecommunications networks for transmitting voice, data and video information. For each multiplexor type there are different power and capacity levels. This is similar to the car engine, which has different cylinder capacities such as 1400 and 1600. The multiplexor is a collection of printed circuit boards (PCBs) – also known as tributaries – which are designed and manufactured in Telco. The PCBs are housed within a sub-rack. The back-plane on the sub-rack is the medium through which all the cards 'talk to each other'. The customer specifies the system requirements for each multiplexor, which is then configured during the manufacturing and installation process.

The PCB assembly process is the most complex sub-assembly process. The number of components can vary quite considerably depending upon the complexity of the design of the board specified by the customer. For example, a PCB may contain as many as 3000 components with over 200 different types. The PCB assembly process may also comprise two PCBs with one layered on top of another. The back-plane assembly process also involves a plug and place operation with electronic components being assembled onto a large PCB. Another major assembly process is the built-in power unit, which is housed in the rack. The customer may also specify the inclusion of a thermal cooling unit in the rack depending upon the climate of the country for installation. The sub-rack and rack assembly processes involve a considerable level of electro-mechanical content. As well as requiring different levels of capacity for transmission products customers are increasingly demanding a wider variety of units to house the equipment. In the past the predominant means of housing the equipment was in European Telecommunications Standards Institute (ETSI) racks for installation in telephone exchanges. However, there is now increased demand coming from newer network operators and the cable television companies. For example, instead of requiring a transmission product in an ETSI standard rack for installation in a telephone exchange, cable television operators are requiring the products to be housed in stand-alone street cabinets.

After the takeover, the organisation set up a design centre near the manufacturing plant for new product development and the redesign of its existing product portfolio. The co-location between the manufacturing facility and the design centre enables design to work hand-in-hand with the product manufacturing operation. The products and systems developed are technically complex requiring, at various stages during the development cycle, the involvement of the disciplines of mechanical and thermal design, electrical and electronic design (digital and analogue), software, systems design, PCB design, manufacturing interface, manufacturing support, component and supply management. The product development life-cycle dictates that many complex interactions take place between these disciplines. The constant changing focus of the world telecommunications markets and pressures on time to market has been creating widespread changes for the design centre. The influences of these changes are being felt directly in the design centre in the form of ever decreasing timescales, pressures to parallel activities and the need to develop more efficient design processes. Technology re-use is used widely within the global company to adapt existing products for specific market applications. The design centre is tasked with the design and implementation of complete systems, and is increasingly influencing the direction that these products take. There are greater demands on the design centre with designers requiring greater knowledge of customer and market requirements in addition to technical expertise.

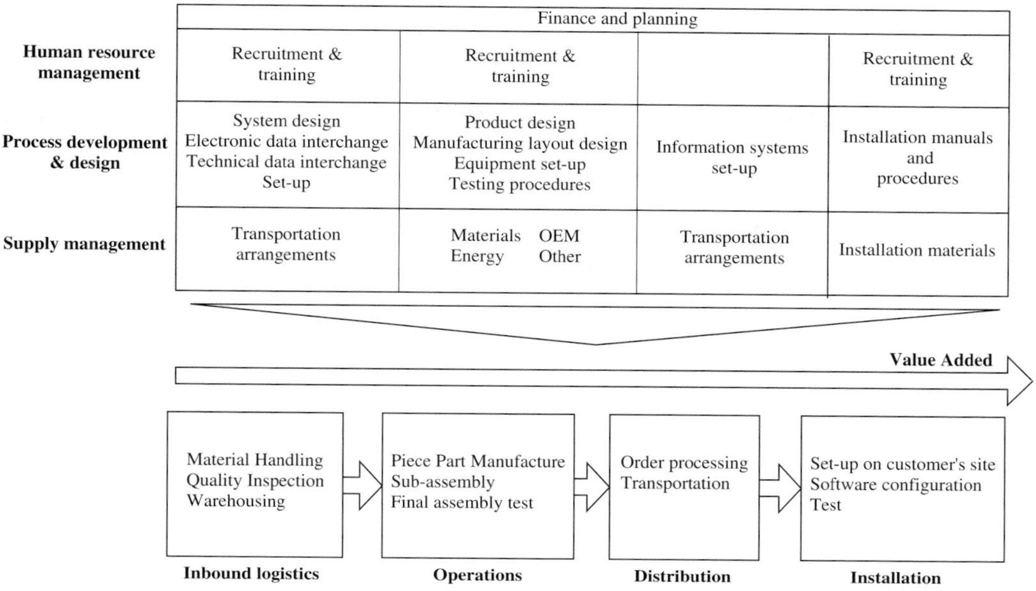

	Finance and planning			
Human resource management	Recruitment & training	Recruitment & training		Recruitment & training
Process development & design	System design Electronic data interchange Technical data interchange Set-up	Product design Manufacturing layout design Equipment set-up Testing procedures	Information systems set-up	Installation manuals and procedures
Supply management	Transportation arrangements	Materials OEM Energy Other	Transportation arrangements	Installation materials

Value Added

Material Handling Quality Inspection Warehousing	Piece Part Manufacture Sub-assembly Final assembly test	Order processing Transportation	Set-up on customer's site Software configuration Test
Inbound logistics	**Operations**	**Distribution**	**Installation**

Figure 11.1 The Telco value chain

Cultivating the relationship between the manufacturing plant and the design centre is an important ingredient in enhancing the competitive position and long-term viability of the site. In the manufacturing site the role of design affects the whole spectrum of product development and manufacturing activities, impacting on the success of projects that range from the re-design of existing products to the development of new products. Therefore, design is a high value-adding activity and increases the technological capability of the site as a whole. Moreover, in Telco the co-location between the design centre and manufacturing is seen as being a significant determinant in ensuring the long-term viability of the site. Building the relationship between the manufacturing plant and the design centre avoids a situation where the site performs the role of what Voss and Blackmon (1996) term a 'screwdriver plant' which carries out assembly processes only with limited design involvement.

Using the value chain provides an illustration of all the key activities that enable Telco to provide the 'total solution' to the final customer. Figure 11.1 illustrates a broad overview of the value chain of activities for Telco.

11.3   Developments in the telecommunications industry

There are a number of significant developments in the telecommunications industry that have been impacting upon Telco and its operations.

- *The liberalisation of service markets* – traditionally, telecommunications was a utility industry similar to power, transportation and the postal service in many countries. Telecommunications was mostly a voice service via telephone with some computer and data communications. Twenty years ago, telecommunications was considered a natural monopoly where the investment in infrastructure was so huge that it was not desirable to allow competition and duplication. This resulted in protected national operators for post, telegraph and telephones (PTTs) – that had high cost structures and poor customer service levels. However, the liberalisation of the telecommunications industries in many countries including Europe and the USA has led to open competition between local and long-distance service providers. The breakthrough in digital technology has dramatically reduced the infrastructure costs and made open competition economically feasible. At the same time, consumer demands have changed from the standard 'fixed line voice only' to mobile telephony, the Internet and proliferation of other services. As a result, many new alternative carriers have entered the market. In fact, significant growth is expected in demand from emerging telecommunications operators such as Energis, Colt, railway companies and electricity companies. For example, railway companies are considering installing their own telecommunications networks by laying cables along railway tracks and then buying the systems equipment. New entrants into the telecommunications market have stimulated demand for a wide variety of systems solutions dependent on the type of service desired and the medium over which it may be carried. This increasing level of competition for the existing service providers has stimulated the need for the large existing operators to upgrade their networks in response. Examples include British Telecom's drive to provide a fully digital network in the UK and modernisation drives being undertaken by the larger service providers in other European countries including Germany.
- *Increasing range of services* – the increasing range of services that can be provided over modern transmission systems has led to opportunities for operators and new entrants to differentiate themselves from the competition. In mature markets, there is unprecedented demand for advanced services, especially for multimedia services that will make use of high capacity, digital broadband networks which form part of the Internet. Niche markets have been opening up to cable companies and mobile services – in turn driving the larger operators to modernise in response to customer demands.
- *Modernisation of networks* – many operators have been striving to modernise networks in response to rapid economic growth that is placing demands on existing infrastructures. The developed PTTs such as British Telecom, Italtel, France Telecom and the Deutsche Bundespost are long established and make up the majority of the total European market. The key market driver for the PTTs is that they already have mature networks. They require modernisation in a cost-effective way that enables them to utilise their current highly

capital-intensive infrastructure. For this reason they require inter-networking and compatibility with existing systems, product solutions based on specific requirements and specifications and suppliers that can deliver time to market at reduced costs.

- *Emerging markets* – emerging markets such as Asia, Africa, Latin America and Eastern Europe are experiencing significant growth. Emerging market nations are building infrastructures as quickly as possible to establish networks that provide not just basic telephony service, but also a full array of advanced services and capabilities, to support rapidly growing industries and business environments.
- *The increasing pace of technological development* – the pace of technological development and levels of integration made possible by developments in the field of micro-electronics has led to the development and installation of new telecommunications networks being more cost effective than retaining older and less efficient systems. Modern backbone and access systems have increased bandwidth and high-speed transmission and switching. This allows the operator to offer and charge for an increasing ranges of services from voice and data to video on demand, home shopping and Internet access. Related to these changes in the telecommunications industry are the innovations occurring in the semiconductor industry. Semiconductor chips are a key constituent of telecommunications equipment and over time, the distinction between different types of chips is beginning to disappear along with reductions in the physical size of the chips (Shillingford, 1998). For example, there are three or four chips in a cellular handset. Chips such as digital signal processors (DSPs) and application-specific integrated circuits (ASICs) are starting to merge with it now being possible to obtain an ASIC for a mobile phone handset with a DSP core built in.

11.4   The impacts on Telco

The influences of these changes in the industry are being felt directly at Telco and the design centre. The nature of the telecommunications industry has changed to such an extent that to continue to compete in a global market, the company has had to be more responsive to change and in particular focus on reducing time to market for its products. The need for cost reduction on existing and new products is driven by the growth in demand for telecommunications infrastructures in emerging markets such as Asia and South America and for cost-reduced versions of existing products. There is also increasing demand from the traditional PTTs for equipment to update their existing networks. In order to reduce costs and remain cost effective in relation to sister plants, Telco pursues a strategy of year-on-year cost reductions across its product portfolio. Product cost is a significant competitive factor in the product portfolio. The company believes that one of the key differentiators for Telco at local level is its ability to achieve considerable cost reductions year on year. Both component cost reduction and re-design require supplier involvement. For example, many of the electronic components included in PCB assembly are undergoing constant innovations and advances in technology that can result in more powerful components at lower costs. There is tremendous focus given to increasing product functionality while reducing the number of PCBs and the number of components on each PCB for the transmission portfolio. Technology re-use is also widely used in the organisation globally to adapt existing products for specific market applications. There is a need to respond to new and fast-moving opportunities in the field of multimedia communications. The requirements of the market are driven by flexible responses based on continually innovating and introducing new services. The globalisation of the telecommunications equipment industry has led many telecommunications equipment makers to move into new markets such as China and South America.

11.5   Strategy formulation at Telco

The strategy of Telco at local level is influenced by corporate level demand for products across all its markets. Particular emphasis is given to the business areas that the corporation wishes to exploit. These plans contain information such as future growth projections, revenues, pricing, market growth rates, new product areas and proposed strategies of the Corporation in each market. At local level the first stage in the process involves Telco carrying out five key reviews.

- *Customer feedback* – information is collected from the annual customer satisfaction survey, regular meetings with customers, and feedback from customer visits.
- *Benchmarking* – is conducted on both a formal and informal basis from Telco's global manufacturing facilities and complemented by visits to other best-in-class companies.
- *Competitive, social and regulatory* – information on their performance is obtained from their customer satisfaction surveys. Additional information relating to social, regulatory and economic indicators is extracted from an analysis of the strategic and operating plans of their chosen market regions.
- *Employee feedback* – annually the company surveys all employees in an attempt to measure their satisfaction levels and help management identify areas for improvement.
- *Suppliers* – annual contract negotiations and monthly reviews are carried out with all key suppliers focusing on a variety of issues. This is a two-way exchange of information where Telco shares details of its strategic and operating plans. Suppliers respond with their view of their business segments, including trends in raw material costs, availability of raw materials, technology movement and obsolescence.

Table 11.2 *Telco's critical processes*

Business function	Critical process
Business planning	Development of strategic/operating plans
Materials	Management of materials supply chain
Manufacturing	Product manufacture/availability
Human resource	Employee satisfaction
Finance	Financial resource management
Customer satisfaction	Customer satisfaction management
Customer service	Product shipment and delivery
Order management	Ease of order input
Engineering	Project management – product transfers

During the development of the strategic plan senior management identify and agree on the critical processes that will enable the company to achieve the objectives of its strategic plan. These critical processes are shown in Table 11.2.

These critical processes must support the key objectives of the company's strategic plan. The key objectives are identified by the senior management team and are aligned with the long-term needs of the business. The key objectives of the company's strategic plan include:

- achieve a 50% reduction in lead time to the customer;
- focus on improving delivery to the customer's required date;
- focus on cost reduction;
- reduce inventory days;
- focus on inter-divisional lead times; and
- reduce overhead as a percentage of standard cost.

The operating plan also details objectives for the key performance metrics.

- *Customer performance* – delivery-to-customer required date, quality commissioning, quality, order lead times, etc.
- *Velocity performance* – order definition to manufacturing, configure and ship, manufacturing velocity, etc.
- *Inventory performance* – days of inventory, inter-divisional lead times, obsolescence management, etc.
- *Financial performance* – cost reduction, overhead as a percentage of standard cost, revenue growth, etc.
- *People* – employee satisfaction, new talent introduction, etc.

Senior management focus on the critical processes that lie at the heart of the company's future success. Senior managers own the critical processes. There is significant focus from corporate level on improving these major processes. This has created an impetus towards linking the company's many locations into common global processes to support global customers. The company must ensure that its critical processes support these global processes. At an operational level, many micro-processes are owned by the team leader and/or employee who is responsible for the operational output of the process on a daily basis. Critical processes are reviewed and improved at two levels:

- *formally* – as part of the business strategy formulation process which result in process changes within the year; and
- *reactionary* – during the year in response to changing business priorities, customer requirements or performance issues.

Each critical process cannot be performed without cross-functional support. For example, one of the key goals of product manufacture/availability is manufacture velocity. This goal cannot be achieved without support from the critical process of management of materials supply chain. The achievement of a critical process depends upon close interaction among functional units and individuals within each function. Furthermore, there must be close interaction between Telco and its suppliers. Due to the fact that the company in some cases is buying in 75% of productive materials for a product, the external supply chain linkages are a crucial part of the equation. The ability of the company to manage and co-ordinate these linkages was viewed as central to achieving its business strategy. A crucial facilitator of this strategy was having a formal policy of determining the best-in-class performers in the industry at carrying out various processes and activities. Information attained from benchmarking served as a key influence upon the formulation of business strategy. Telco considers the relative performance of sister plants and companies in which the corporate organisation has a financial stake along with its external competitors. There is considerable internal competition between each manufacturing site in an attempt to 'impress' corporate level. By 'impressing' corporate level and performing favourably in comparison to sister plants Telco at local level enhances its prospects of being chosen as a plant for the introduction of new products. As well as benchmarking its external competitors in the industry, the company also benchmarks companies in other industries. For example, the company recently benchmarked a catalogue retailer in the area of order processing and bar-coding. The benchmarking process involved Telco people visiting the catalogue retailer and identifying the operational and managerial processes of order processing. It then selected the comparable processes to Telco's business and assessed the performance of these processes in relation to cost, quality and time.

11.6   Outsourcing evaluation and management cases at Telco

This section presents an overview of three case studies illustrating the issues involved in evaluating and managing the suitability of a number of activities for outsourcing. The issues involved in these three case studies are then related to the stages on the outsourcing framework outlined in this book.

11.6.1 Case one

In the telecommunications and electronics industries many organisations have been using logistics service providers (LSPs) to manage inventories in order to reduce time-to-market and reduce costs. Many original equipment manufacturers (OEMs) have been purchasing up to half of their component requirements through logistics service providers, rather than directly from manufacturers. Logistics service providers have become key sales centres for OEMs while the component manufacturers have become high volume suppliers for the logistics service providers. This case study outlines the implementation of a trading arrangement between a logistics service provider and Telco. Implementing this arrangement involved Telco outsourcing many of the activities associated with the logistics and procurement processes. Due to the nature of the assembly process Telco had been purchasing a wide variety of electronic components from a large number of suppliers. In the past, the buyers in the purchasing function spent a considerable amount of their time in transaction type activities such as order planning, expediting and resolving supply problems for many of these components, with a lot of this time being spent on low-value commodity type components. Although Telco had implemented electronic data interchange (EDI) successfully with its key suppliers, it was not running EDI effectively with the majority of its commodity type suppliers. For example, in many cases the delivery forecast was sent by fax. Firm orders were being sent by EDI with the buyer in Telco also sending a fax of the same order as well. In the supplier organisation inefficiencies were also occurring with the supplier not dynamically loading the EDI message onto their manufacturing systems but printing the message out and manually entering it into their systems. The buyers in Telco were responsible for managing the suppliers while Goods Inwards managed the transfer of delivered items to stock.

As a result of these problems, Telco decided to use an electronic component logistics service provider to manage the logistics process between 50 of their commodity type suppliers. The objective was to ensure comparable service to that of internal functions, whilst increasing efficiency and reducing transaction costs with these commodity suppliers. In effect, the logistics service provider would act as the interface between the Telco site and each component supplier maintaining and managing a store of items on the site. The operation of the entire system depends upon electronic communication with limited intervention in Telco. The logistics service provider uses its distribution centre as the 'hub' for the items it purchases from the suppliers. The initial running of the system begins with the logistics service provider receiving a forecast of usage for each component for the proceeding 6 months. Consequently, each week the logistics service provider receives a revised forecast of usage for the next 4 weeks from the manufacturing system of Telco. Based on this forecast the logistics service provider must ensure that its on-site store has a stockholding equivalent to this forecasted 4-week usage. On receipt of the weekly forecast the logistics service provider guarantees 24-hour delivery to ensure necessary replenishment of the on-site store. The logistics service provider still 'owns' the stock while it is in the on-site store. When Telco's manufacturing systems uses the stock from the store it then becomes the property of Telco. Figure 11.2 shows the set up of the system between Telco, the logistics service provider and the component suppliers.

The LSP is operating a consigned stock arrangement with Telco. At Telco's site, the LSP maintains a stock equivalent to 4 weeks usage in Telco based on a weekly forecast transmitted electronically. At the LSP's distribution centre, additional stock and purchase order coverage is provided in order to meet the manufacturing requirements of Telco. Inventory levels will be at a level equating to a minimum 10% of forecasted annual usage. In relation to payment and component issue, manufacturing in Telco draws components as required from the LSP's store and automatically updates the stock balance held on the manufacturing system. Each week Telco electronically notifies the LSP of all components drawn from its component carousel in Telco by part number, value and the date drawn. At the end of each month the LSP issues an invoice electronically giving a breakdown of the usage of each component for that month of issue to Telco with payment due in 30 days. This effectively is 'self-billing' driven by usage, with weekly matching and monthly payment. Price is the purchase price from the original supplier of the components and includes a percentage mark-up agreed between Telco and the LSP. In implementing this arrangement Telco has drastically altered the business processes associated with purchasing to the point where it is no longer carrying out its traditional role. The major benefits that the LSP offers Telco under this arrangement are the following.

- The LSP is providing 'local' inventory with Telco paying on use.
- Telco is obtaining reduced lead times and quality inspection for the components that have become the responsibility of the LSP.
- Rationalisation of the supply base and a reduction in transaction costs.
- The implementation of Internet technologies requires less purchasing resource with the buyers no longer being a link between the company and suppliers with Internet technologies making the processing of requisitions, order acknowledgements and invoices redundant.
- The buyers affected by this arrangement are able to focus on more value-adding tasks in the company such as implementing early supplier involvement in new product development activities.
- Telco is using an external source that can provide greater levels of expertise and service in managing the logistics process than Telco would achieve internally.

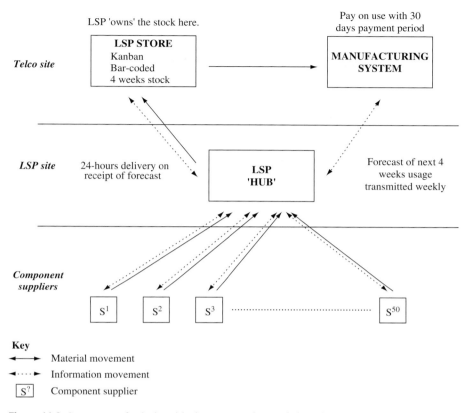

Figure 11.2 Structure of relationship between Telco and the LSP

11.6.2 Case two

This case is concerned with the outsourcing of a key electro-mechanical sub-assembly process. The electro-mechanical manufacturing processes have become characterised by a high percentage of activities being outsourced to firms manufacturing systems and components. In many cases this percentage can be greater than 80%. The sub-rack is the means by which the PCBs that comprise the multiplexor are housed. Telco assembles a variety of sub-racks for each multiplexor. Traditionally, the assembly of the sub-rack was carried out internally within Telco with various piece parts and sub-assemblies being sourced from a number of suppliers. The two key sections of the sub-rack are listed.

(1) Back-plane assembly – the back-plane is a large PCB, which acts as the medium connecting the PCBs in the sub-rack. The back-plane assembly process is carried out in-house involving a plug and place operation with electronic components such as resistors, connectors and cable assemblies.

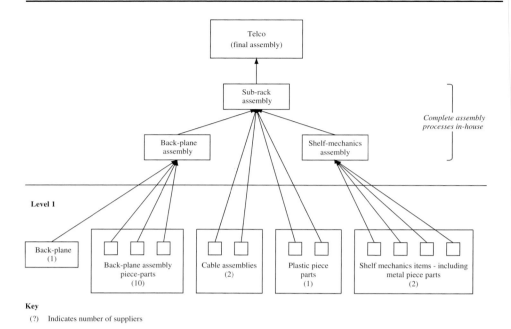

Key
(?)   Indicates number of suppliers

Figure 11.3 Organisation of supply base for in-house sub-rack assembly

(2) Shelf mechanics assembly – the metal piece parts that comprise the fabrica-
tion of the sub-rack. The majority of the piece parts (80%) for the shelf
mechanics are sourced from a sub-assembler with some of the piece parts,
including aluminium side panels sourced from a specialist metal supplier.
This sub-assembler is a UK-owned engineering company specialising in
supplying the telecommunications industry and is one Telco's key sub-
assemblers.

Figure 11.3 shows the organisation of the supply base for in-house assembly.

In the design process for a new multiplexor, Telco decided to re-design the sub-
rack to facilitate the outsourcing of the assembly process to the sub-assembler.
With considerable interaction from the sub-assembler, Telco moved towards the
design of an all-steel sub-rack. This would enable the sub-assembler to carry out
the majority of the metal manufacture in-house. The sub-assembler would buy in
the other metal and plastic piece parts as well as the assembled back-plane from
the chosen back-plane supplier. Figure 11.4 shows the organisation of the supply
base under this new arrangement. A crucial factor in choosing this particular sub-
assembler was the fact that it had recently opened a plant in a nearby town and was
able to provide Telco with an efficient and reliable delivery service.

The reasons for Telco moving towards this arrangement were:

- reduced cost of the sub-rack;
- reduction in the number of direct suppliers;

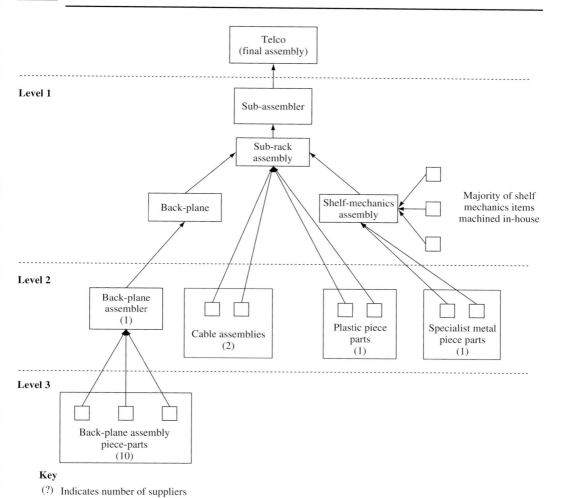

Figure 11.4 Organisation of supply base for outsourced sub-rack assembly

- less purchasing resource required due to supply base reduction;
- the responsibility for the management of inventory, logistics and quality moved from Telco to the sub-assembler; and
- potential to further strengthen the relationship with one of its key sub-assemblers.

11.6.3 Case three

This case relates to the evaluation of the design capability of Telco for an electronic component – an application specific integrated circuit (ASIC). This component is a key driver in the functionality of each PCB associated with the multiplexor. There are various components associated with each PCB that give the multiplexor its functionality. In relation to the components that comprise the PCB assembly there are two types.

- *Generic* – these are defined by designers as simple and standardised components that can be readily sourced from a number of suppliers. Examples of such components include resistors or relays. Although these items may be 'simple and standardised' they tend to be purchased in high volumes. Items such as capacitors or resistors are relatively inexpensive. However, one PCB assembly may have as many as 200 resistors of one type with a similar number for some types of capacitors.
- *Specific* – these are key components in the assembly of the PCB that can have a considerable impact on the power and functionality of the multiplexor as well as accounting for a considerable cost. Examples of such components include ASICs, field programmable gate arrays (FPGAs) and microprocessors. Although there may be only a few *specific* components on each PCB, each of these components can account for over 15% cost of the total PCB cost. Advances in the design of these components can lead to the development of a more functional and lower cost product that can be a potential source of differentiation in the marketplace.

Due to increasing customer demand and a high level of competition in the market-place, there has been a constant emphasis in Telco on achieving advances in the design of these specific components. These developments have been placing considerable pressures on design resource within the design centre. Management in the design centre believed that a more focused approach should be adopted to allocating design resource in areas that could deliver long-term value for the company. At the same time, suppliers were becoming more competent in the design of these components and were actively encouraging their customers to allow them to design as well as manufacture them. In fact, some of the competitors of Telco had already outsourced some of the design for ASICs. However, by divesting its capability in this area, the company could potentially lose a source of competitive differentiation both currently and in the future, and this danger was further exacerbated by over dependence on a limited number of suppliers. Telco had carried out extensive analysis of its own capabilities in ASICs design relative to its suppliers and competitors. The analysis revealed that it was marginally more capable than most of its competitors and suppliers. However, the analysis had revealed that two of its direct competitors had been investing in this area over the last few years. Consequently, Telco had to make a decision on whether to rely more heavily on suppliers or invest in this area in order to build upon its current capabilities.

In this case, Telco decided to keep this design process in-house and invest more resource in this area in order to build upon its current capabilities. The company adopted a more focused approach to design that involved outsourcing the design of less critical processes to suppliers. In particular, the company decided to re-allocate the resource from a number of electro-mechanical design areas to the design of specific components. Telco also believed that some of its key electro-mechanical suppliers had stronger design capabilities than in-house. In relation to the strategy for maintaining and developing its capabilities in ASIC design, Telco placed heavy emphasis on better accessing the capabilities of suppliers. There had already been a high level of supplier involvement in the design process for these components. For example, suppliers had been involved at the concept stage in the design process providing prices and design suggestions. However, it was felt that a more formal and structured approach with a limited number of suppliers could deliver more benefits for Telco. Although the design teams in Telco had been designing the component, the design process was heavily influenced by the current and future capabilities of the suppliers in the technology. For example, designers might design an ASIC and arrange for its manufacture by one of its key ASIC suppliers. The company uses ASICs either because there is no standard solution or because a standard chip would not give them the functionality or level of power consumption required. Due to the high costs and risk associated with their design and manufacture a higher level of interaction is required between the designers and the chosen ASIC manufacturer. Technology roadmaps that outline the evolution of the supplier's capability in a technology are a crucial part of the interaction process between the design function and the supplier. Effective supplier involvement in this process is crucial due to the rapid changes occurring in these technologies and the constant focus on re-design for cost reduction and greater component functionality throughout the life of the product.

11.7   Relating the cases to the outsourcing framework

Case 1 – the use of a logistics service provider

Stage 1   activity analysis
Activities:
- inbound logistics (including material handling, quality inspection and warehousing) and procurement.
  Supplier roles required:
- provide local inventory with Telco paying on use;
- provide demand-pull delivery;
- ship - to - point of use;
- maintain and manage on-site store; and
- manage component obsolescence.

Stage 2   importance level analysis

Defined as a 'critical' activity for the following reasons:

- high levels of performance in this activity have a major impact upon the key success factors for Telco including cost reduction and inventory reduction; and
- contributed to reduced time to market for customers, which had become a key competitive differentiator in Telco's product markets.

Stage 3   relative capability analysis

LSP considered to be more capable for the following reasons:

- the LSP could provide a higher level of service at a lower cost than Telco could achieve by continuing to carry out the logistics and procurement activities internally for this portfolio of components; and
- although Telco was paying a commission to the LSP for the arrangement, the LSP was providing components at a much lower price because of its volume buying from suppliers for a number of customers.

N.B. In comparison with the traditional arrangement with its suppliers, Telco believed it would reduce its costs through better inventory management (via consigned stock, ship to point-of-use and reduced lead times) and less procurement resource involved in managing suppliers. From the perspective of Telco, the entire system would be managed electronically with limited procurement involvement.

Stage 4   influences on strategic sourcing options

The major influences on the decision to 'strategic outsource' were as follows:

- Telco was not in a position to replicate the capabilities of the LSP internally;
- competitors were increasingly using LSPs to realise the potential benefits;
- risk was reduced with component obsolescence becoming the responsibility of the LSP;
- although there was some resistance internally, Telco believed it could manage this through the re-deployment of the affected staff; and
- supply market risk was deemed manageable because there were a number of other capable LSPs available if the outsourcing process with the chosen LSP failed to deliver the required benefits.

Stage 5   relationship strategy

The objectives for the outsourcing process included the following:

- to ensure comparable service to that of logistics management and procurement, while minimising the total cost to Telco;

- cost reduction in the form of lower priced components and less logistics and procurement resource;
- inventory reduction;
- reduce the risks associated with component obsolescence;
- demand-pull delivery from the LSP; and
- rationalisation of the supply base.

The 'competitive collaborative' relationship was chosen for the following reasons:

- Telco initially wanted to assess the performance and strength of the relationship with the LSP before committing to a 'close collaborative' relationship; and
- the presence of three other potential LSPs facilitated the employment of this strategy.

Stage 6   establish, manage and evaluate appropriate relationship

- Considerable interaction between Telco and the LSP concentrated on determining how the structure of the arrangement would operate. Issues addressed included how the LSP would manage the on-site store, bar coding, ease of interface with Telco systems, and the billing process.
- Initially, the principal concern was determining how many components and number of suppliers to include in the process.
- It was decided to apply it to 200 components from 50 suppliers in order to evaluate its success and the feasibility of further application to other components in its portfolio in the future.

Case 2 – the sub-rack assembly

Stage 1   activity analysis

Activities:

- inbound logistics including material handling, quality inspection and warehousing;
- operations including piece-part manufacture and sub-assembly; and
- procurement.

Supplier roles:

- inventory and logistics management;
- provide demand-pull delivery schedule;
- chosen supplier would manage lower tier suppliers; and
- assembly and design capability.

Stage 2 importance level analysis
Defined as a 'non-critical' activity for the following reasons:
- sub-assembly activities were not considered to be a source of competitive differentiation in the marketplace;
- Telco wanted to focus on more value-adding activities; and
- there were a number of competent suppliers that could provide these sub-assemblies at a lower cost.

Stage 3 relative capability analysis
The analysis focused on the following:
- analysis of costs of materials for an all-steel sub-rack in comparison to original design;
- analysis of Telco assembly costs in relation to the sub-assembler; and
- an analysis of the current and historical performance on a range of criteria of the sub-assembler's business with Telco.

The sub-assembler was considered to be more capable for the following reasons:
- possessed a considerably lower cost base than Telco due to economies of scale and lower labour rates;
- the sub-assembler focused on sub-rack manufacture and also possessed design capabilities; and
- the sub-assembler was willing to manage the lower tier suppliers and providing demand-pull delivery to Telco.

Stage 4 influences on strategic sourcing options
The major influences on the decision to 'outsource' were as follows:
- Telco could not achieve the cost position of the sub-assembler;
- the sub-assembler had established a site close to Telco;
- supply market risk manageable because there were a number of available sub-assemblers willing to take on the business; and
- although there was some resistance internally, Telco believed it could manage these difficulties through the re-deployment of the affected staff both internally and to the sub-assembler.

Stage 5 relationship strategy
The objectives for the outsourcing process included the following:
- achieve a significantly lower priced sub-rack;
- reduction of 16 suppliers to one;

- inventory reduction; and
- demand-pull delivery from the sub-assembler.
    'Adversarial' relationship chosen for the following reason:
- initially, Telco classified this relationship as 'adversarial' because of the high level of competition in the supply market. However, if the outsourcing process proved successful, then it would consider outsourcing more sub-assembly processes and further strengthening the relationship.

Stage 6   establish, manage and evaluate appropriate relationship
- Redeployment of staff and equipment from Telco to the sub-assembler; and
- the sub-assembler had recently opened a plant in a neighbouring town, which had the potential to strengthen the relationship.

Case 3 – ASICs design

Stage 1   activity analysis
Activities:
- design
    Supplier roles:
- software and hardware design capability;
- prototyping; and
- testing facility.

Stage 2   importance level analysis
Defined as a 'critical' activity for the following reasons:
- source of competitive differentiation in the marketplace;
- a number of competitors of Telco had become more competent in the technology through investment; and
- any innovations developed in this technology could be exploited in a number of Telco's products.

Stage 3   relative capability analysis
This analysis had revealed that Telco was more competent than its suppliers and many of its competitors. However, two of its competitors had an advantage over Telco in terms of technology breakthroughs, which create higher functionality and lead to cost reduction.

Stage 4   influences on strategic sourcing options

The major influences on the decision to 'perform internally and develop' were as follows:

- a strong position in this technology was perceived as a source of competitive differentiation;
- Telco already had some capabilities in this technology, which had shown potential for development;
- by re-allocating investment from less critical design areas, it was possible to strengthen the position of Telco in this technology;
- also, by building closer relationships with suppliers in this technology, the capability of Telco could be strengthened; and
- outsourcing such a capability would have been a major psychological blow to designers within Telco.

11.8   Discussion

The three cases have illustrated the influences on the choice of each sourcing option. The three case studies are illustrated on the strategic sourcing options matrix on Figure 11.5. In *Case* 1, the major driver for outsourcing was the capability of the logistics service provider to provide a higher level of service at a lower cost than internally. Telco was not in a position to replicate this capability. Although the logistics activity was critical to the success of Telco, the company believed that outsourcing the activity would enhance its competitive position, and the company considered the level of supply market risk as manageable. In *Case* 2, the major influence was the relative cost position of the supplier in the assembly process. Moreover, outsourcing this type of activity was central to the company's strategy of moving away from standard sub-assembly processes in order to focus on more value-adding activities in which it was in a better position to establish a superior performance position. Telco reduced the level of internal resistance through re-deploying the affected staff both internally and with the sub-assembler. In *Case* 3, the dominant influence on keeping ASICs design internally was due its importance to the competitive position of the company. Although there were a number of competitors more competent in the design process, the company believed it would enhance its own position through additional investment in this area. ASICs design was also perceived as a critical area for the future development of the organisation. However, as evidenced in *Case* 1 and *Case* 2, there has been a significant trend by the company towards outsourcing activities within its value chain. It is clear that the company was concentrating on the final assembly sub-activity while suppliers were increasingly carrying out the sub-activities of

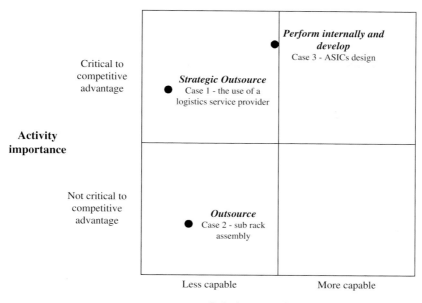

Figure 11.5 The strategic sourcing options matrix

piece-part manufacture, sub-assembly and test. The outsourcing of sub-assembly processes has implications for the inbound logistics activity of the value chain. The sub-assemblers were responsible for co-ordinating and procuring inbound piece-parts from lower level suppliers – activities formerly carried out by Telco. This illustrates the emphasis being placed on the skills of suppliers in inventory and logistics management in the supplier selection decision. Suppliers were providing an increased level of service in the form of more frequent deliveries and increase in production planning activities. In *Case* 1 the supplier was almost taking on full responsibility for managing the inbound logistics and procurements activities. There was limited need for procurement interference with the buyer–supplier arrangement being managed electronically. These examples illustrate the evolution of the shifting responsibilities for carrying out activities between Telco and its supply base. The trend towards outsourcing by the company can be attributed to the following reasons.

- *Most competent source* – the company's outsourcing policy was based upon the best available source (internal or external) being chosen to carry out the activity or group of activities. By using the most competent source Telco argues it was positively impacting upon its strategic objectives.
- *Increased flexibility* – Telco believed it could be more flexible by outsourcing more activities rather than performing activities internally by being in a better position to react rapidly to market changes and be more responsive to customer change.

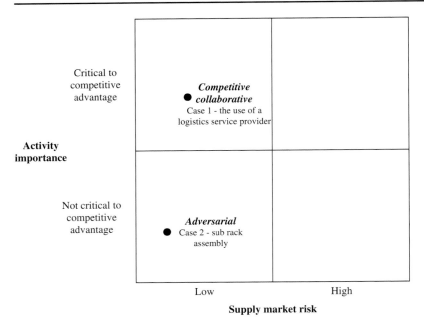

Figure 11.6 The relationship strategy matrix

- *Reduced risk exposure* – through outsourcing the company was reducing its level of risk. For example, by gradually outsourcing manufacturing processes it was reducing risk by converting its fixed costs into variable costs. In times of adverse business conditions suppliers would then have to deal with the problem of excess capacity. The company argued that suppliers were better able to cope with demand fluctuations through economies of scale and had more scope for alternative sources for this excess capacity.
- *Cost reduction* – in some cases the activity could be performed at a lower cost by outside suppliers. For example, the company believed it was reducing costs by outsourcing sub-assembly activities by availing of suppliers with lower wage structures in comparison with the company. Suppliers could also offer lower costs through economies of scale. The company was further reducing its costs through using less purchasing resource and promoting 'the re-allocation of the buyer's efforts towards 'new product introduction'.
- *Supplier management* – approximately 80% of purchases (by value) were obtained from 41 suppliers that the company defined as key suppliers. Telco believed it was reducing the level of risk associated with such a high bought-in content by employing effective supplier management and partnership building approaches.

Figure 11.6 shows the relationship strategy matrix for each of the outsourced activities. Although a *close collaborative* relationship strategy was not adopted initially in either *Case* 1 or *Case* 2, the company considered that once the relationships had developed with the chosen suppliers over time, there was the potential to develop higher levels of collaboration. In particular, in *Case* 1, if the arrangement proved successful then the company intended to source a larger portfolio of components from the LSP, which in turn would involve further strengthening the relationship. The experiences of Telco emphasise the importance that the supply market can have upon the outsourcing process. In fact, analysis and use of the supply market can have a more significant influence on the outsourcing process rather than attempting to identify the importance of the activity. For example, *Case* 1 illustrates a situation where the company was using a supplier far more competent in the inbound logistics activity of the value chain. By using a more competent external source, the company was raising standards in the inbound logistics activity as well as contributing directly to business strategy objectives. The company required a supplier that would provide local inventory and JIT delivery. Therefore, the company chose a supplier that specialises in this area of logistics. High performance in the logistics activity was central to the success of the company and a potential source of differentiation in the marketplace. In *Case* 2, another key influence on outsourcing the sub-assembly process was the fact that the chosen supplier had located near to Telco and was in a position to offer higher service levels. Telco had been attempting to develop more collaborative relationships with a number of its key suppliers in order to enable the outsourcing of a number of other business activities. Although company personnel argued that collaborative relationships with a number of suppliers were being developed, it was found that in some instances this certainly was not the case. This was particularly evident in the following areas.

11.8.1　Joint buyer–supplier cost reduction

The company pursued a strategy of year-on-year formal cost reductions across its product portfolio. Although cost reduction may have focused mainly on internal operations there was significant emphasis on attempting to realise cost reductions in the supply chain. Consequently, this had implications for suppliers and the role of the purchasing function in managing the role of suppliers in the cost reduction process. In particular, it was clear that a change in the traditional attitudes of the people involved in managing the relationship was required in relation to cost reduction in the supply chain in the areas of supplier switching and joint buyer–supplier cost analysis.

11.8.2　Supplier switching

There were a number of instances to illustrate a lack of understanding of the true meaning of collaborative buyer–supplier relationships and a reluctance to acknowledge that change that had taken place in the way that the company conducted its business. Furthermore, there was a lack of understanding of the implications of the changes for the individual. For example, it was evident that buyers were still wedded to the belief that the threat of switching supply sources was an entirely justifiable method for achieving cost reductions while at the same time expounding the virtues of collaborative buyer–supplier relationships. There were also instances of the buyer's attitude to cost reduction in other areas of supplier management. For example, in implementing supply base reduction strategies across a variety of commodities buyers were not convinced of the benefits of moving towards a reduced supply base believing it to increase risk exposure and to provide less opportunity to switch suppliers to achieve cost reductions. With electronic components undergoing constant innovations in terms of both functionality and cost, this provided problems and conflicts in the cost reduction team. For example, it was quite common for an existing component that was being supplied to become uncompetitive if another supplier introduced a much more advanced component to the market. This scenario led to the design and cost functions proposing switching suppliers. However, such a move provided problems for purchasing as it conflicted with their objectives of developing more collaborative relations with the same suppliers.

The experiences of Telco signal a lack of understanding, trust and commitment with regard to the concept of true 'collaboration' between buyers and suppliers. It would seem fair to argue that the company may have decided at a strategic level to pursue collaborative buyer–supplier relations. The evidence presented would suggest that this strategic decision was supported by superficial agreement amongst organisational members that closer collaboration between buyers and suppliers would be mutually advantageous. However, the behaviour of those involved in cost reduction activities would indicate that little thought was given to the task of identifying the necessary systemic changes that must be effected in order to achieve greater levels of collaboration.

Limited level of joint buyer supplier cost reduction

The company operated open book arrangements with its key suppliers having access to cost information such as material costs, packaging and delivery costs, overheads and profits. There was limited evidence of the company working jointly with suppliers to measure the total cost of ownership – costs associated with the acquisition, use and maintenance of a good or service throughout the entire supply chain (Ellram, 1996). The purchasing manager in the company agreed that this was a major impediment to more collaborative supplier relationships. For example, during price negotiations with key suppliers he was 'very sceptical' of cost breakdowns provided by suppliers to justify any component price changes. This also made it very difficult to constructively negotiate prices with limited information on supplier internal operations without being seen to be creating a win/lose situation. When asked why there was a low level of joint customer–supplier cost reduction activity the purchasing manager identified the major problem to be that of limited resource in the purchasing function. Although the purchasing manager had been successful in securing resource for recruiting engineers into the purchasing function he had been less successful in convincing senior management of the benefits of recruiting a financial analyst to pursue joint cost analysis with suppliers.

11.8.3 Supplier integration in new product development

As part of the attempts to develop more collaborative supplier relations the company had been attempting to increase the level of supplier involvement in their new product development activities. However, there was evidence of the traditional attitudes of members of the cross-functional team hindering the increased involvement of key suppliers in new product development activities. For example, there were inconsistencies in the management of its key suppliers in new product development along these dimensions. Although there was a high level of collaboration between the company and its suppliers in the design process, in the areas of information sharing, there was conflicting evidence on the nature of involvement of suppliers in the design process. In some instances, the supplier was heavily involved and selected at the concept stage of the design process – a key electronic component supplier. In other instances, the role of the key supplier was limited to that of providing information on prices and lead times with the supplier selection decision being made at the development stage. There were a number of barriers identified to increasing supplier involvement in new product development including the following.

- Attempting to increase the role of the supplier in the design process led to resistance from designers perceiving it as a threat to their employment if the company outsources design activities.
- The company was not accustomed to involving suppliers at an early stage of the design process. The traditional way of dealing with suppliers was that of requesting information, not accessing the skills and knowledge of the supplier into the design of the product. There was still a perception that suppliers could be played off against one another in order to extract more favourable terms such as price and lead times.

11.8.4 Delivery and logistics management

The importance of delivery and logistics management was borne out by the fact that the following key objectives of the company's business strategy were directly related to the effective management and integration of supplier materials into the manufacturing process including achieving a 50% reduction in lead time to the customer; reducing inventory days (including pre-process, work-in-progress and finished goods); and focusing on inter-divisional lead times. The policy the company had been pursuing was to extract a higher level of service from its suppliers in the form of shorter lead times and more responsive and flexible delivery schedules. At the same time, the company was providing erratic demand forecasts to many of its suppliers. The instability of the forecast was felt most starkly amongst the company's sub-assemblers that were managing the lower tier suppliers. A great deal of emphasis was placed on the organisational and logistics skills of the sub-assemblers being responsible for co-ordinating the in-bound materials from the lower level suppliers that go into the sub-assembly processes. Analysis had revealed that the key issues from both the supplier and customer sides were the following.

- *Supply side* – on the supplier side, the most common concern was sudden upsurges in the customer's demand schedules leading to a safety stock being held on the supply side. Another related issue to arise was that the customer was using suppliers to hold inventory to allow for poor forecasting on the customer side. Therefore, suppliers were suspicious of the motives of the customer when the customer outsourced assembly processes, which led to suppliers being responsible for co-ordinating materials from suppliers at lower levels in the supply chain. Suppliers were wary of being responsible for 'managing' suppliers. The supplier perception of delivery and logistics management was that of the customer 'shifting' inventory responsibility to the supplier and the supplier becoming liable for any supplier failure at lower levels in the supply chain if there were any sudden upsurges in demand from the customer.

- *Customer side* – on the customer side, the company agreed that more adequate forecasts of future requirements and stable delivery requirements should be provided to suppliers. This would be improved through better communication with marketing, materials management, purchasing and manufacturing. However, a considerable barrier identified by the purchasing manager was that of limited resource allocation from the materials management function with senior management not displaying a great degree of urgency in alleviating the problem of the safety stock on the supply side. This was not surprising given that the company had been extracting higher levels of service from suppliers without providing any additional remuneration. Such attitudes on the part of senior management reveal how the customer is attempting to improve its own efficiency by shifting the problem of poor forecasting lower down the supply chain.

The creation of a flatter and more responsive organisation is considered to be one of the advantages associated with outsourcing (Quinn and Himler, 1994). However, the experiences of Telco have shown that this can be quite difficult to

achieve due to the traditional functional view of the organisation held by personnel. In most organisations, there are natural reasons for the reluctance of internal
functions to collaborate and managers are not usually trained or developed as
collaborators (Beardwell and Holden, 1997). They usually progress across hierarchical levels by performing duties for which they are directly accountable. The
implications of a lack of internal collaboration are that the company is going to
have considerable difficulties adopting collaborative arrangements with external
suppliers if it cannot develop a 'partnering' mindset across internal functions.
The prevailing culture in both the customer and supplier organisation should
facilitate and encourage joint problem solving and decision making across organisational boundaries. The experiences of Telco have stressed the need for a
culture that fosters a mindset, views and beliefs that acknowledge inter- and
intra-organisational collaboration, moving away from the traditional functional
view of the firm.

The findings have also revealed that there are considerable impediments for
those participants responsible for establishing and managing more collaborative
buyer–supplier relationships. These impediments exist at both the strategic and
operational level. At the strategic level this was evident in the following areas.

- Successful collaborative buyer–supplier relations depend upon a high level of
  commitment and resource allocation from both the customer and supplier
  organisations (Ellram and Edis, 1996). In Telco, senior management were
  reluctant to allocate sufficient resource in the areas of joint buyer–supplier
  cost analysis and delivery management to support more collaborative relations.
- Ramsay (1996) argues that collaborative buyer–supplier relations depend upon
  the buyer being in possession of a very large E/R ratio, that is the ratio of the
  buyer's expenditure to the supplier's total sales revenue. The objective is to
  create a high level of mutual dependency between the partners. However, within
  Telco, the policy of senior management was that none of their key suppliers
  should have an E/R ratio higher than 30%. Such a policy was based on the belief
  of the senior management team that there was too much risk associated with
  giving one supplier such a high percentage of business. However, such a policy
  has a self-defeating influence on the whole ethos of collaboration by reducing
  the likelihood of high mutual dependency between the partners.

At the operational level the following impediments existed.

- Members of the new product development team were not consistent in the
  management of key suppliers in the new product development process.
  Designers were opposed to increasing the involvement of suppliers perceiving
  it as a threat to their employment.
- The buyers were not convinced of the benefits of moving to a reduced supply
  base fearing that it increased risk exposure and limited the opportunity for cost
  reduction in the supply chain by not being able to switch supply sources easily.

These findings have also shown that it is not enough to change the attitudes of the purchasing personnel but the attitudes of the other business functions and senior management must also be changed. Effective implementation of collaborative buyer–supplier relations requires a culture permeating the organisation hierarchy that encourages and values collaboration. However, such a requirement presents organisations with an immense challenge due to the embedded culture of both the buyer and supplier who have traditionally operated on an arm's length basis.

11.9 Going modular at Telco

In relation to Telco, the site had 400 productive material suppliers with the intention of reducing the number to 200 within a 3-year time period. This was considered to be a radical transition in the size and organisation of its supply base. However, a question that arises is how far the company is prepared to go with outsourcing and supply base reduction. Figure 11.7 presents an overview of the intended organisation of the major parts of the supply base and an approximate supplier count for each. The major assembly process that the company intended to keep in-house (unless there were any capacity problems) was PCB assembly. The company's sub-assemblers carried out all the other major sub-assembly processes. In the analysis of Telco's business strategy it was shown that the skills associated with integrating the services and outputs of activities had become more important as the company moved from a strategy based on tangible assets to a strategy based on intangible assets and capabilities. In effect the company was gradually becoming a 'systems integrator' implying that all manufacturing will be done externally by suppliers.

It is worth illustrating how Telco and its key suppliers would be organised under a modular production arrangement. Figure 11.8 illustrates how Telco's and a number of its key suppliers would be organised under a modular production arrangement with all the major manufacturing and assembly processes carried out on a single site. With all the major manufacturing and assembly processes taking place on a single site it appears similar to a vertically integrated arrangement. However, the distinct difference is that the customer is achieving vertical integration without financial ownership. Under this scenario one supplier would be responsible for each of the major parts of the customer's purchasing portfolio. Each supplier is a 'partner' taking on greater responsibilities in return for a longer-term commitment from the customer. Each supplier is responsible for the 'supply' of all items included in each supplier category. This responsibility would involve manufacturing items as well as managing the material flow from any specialist

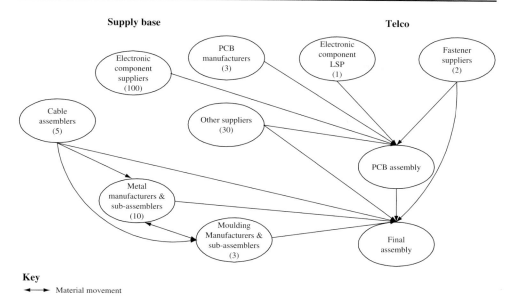

Figure 11.7 An overview of the major parts of Telco's supply base

suppliers in each category. The customer would be responsible for the final assembly process only with all major sub-assembly processes carried out by suppliers. Suppliers would have a more prominent role in product design, production engineering and involvement in layout and workflow in final assembly operations. In effect, each partner is part of a network of specialists focusing on their distinct area of competence delivering products and services to the systems integrator. Suppliers would be located as close as possible to the final assembly point in the customer organisation. Rather than being located in adjoining supplier parks, the customer and all the major suppliers would be located on a single site. The Logistics Service Provider would assume responsibility for managing the supply of all electronic components to the appropriate partners on the site while designers in the customer would still maintain its links with the electronic component suppliers.

Supplier integration under modular production has the potential to promote greater collaboration in a number of areas.

- *Supplier involvement in design* – such an arrangement would enable suppliers to have an increased role in design with the systems integrator. The customer would retain control of the key areas of product design such as specific electronic component and software design, which have considerable impact on the power and functionality of the products. These areas are perceived by the customer as adding value and enable the company to maintain its competitive position. However, suppliers would take a more active design role in areas such

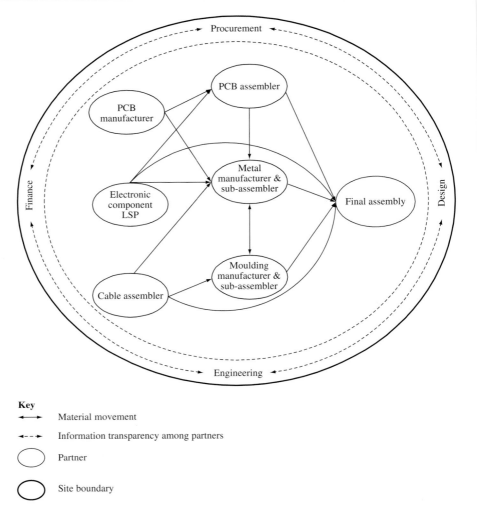

Figure 11.8 Modular production and supplier integration in Telco and its key suppliers

as metal and plastic design. The increased role of suppliers is facilitated by the proximity to the customer and therefore increased information exchange.

- *Joint cost reduction* – one of the major barriers to joint cost reduction between buyers and suppliers is the diversity of manufacturing cost accounting systems in the supply chain. However, with only a limited number of suppliers it is more appropriate for each supplier to use the same accounting systems as the customer. Suspicions of the customer using supplier costing information for spurious reasons would be limited because of the high mutual dependency of each partner.
- *Delivery practice* – under the modular arrangement piece-part manufacture is happening as close to the point in time and location it is needed rather than

bringing in a large sub-assembly from suppliers in geographically dispersed locations. This would involve suppliers working within the customer facility and
manufacturing and assembling piece parts that are fed directly to final assembly.

- *Active information exchange* – the closer the supplier is to the customer the more
likely there is a greater intensity of information exchange. The 'single site'
concept facilitates immediate communication in the event of any problems in
areas such as quality. One of the major barriers to electronic information
exchange between customers and their suppliers was the incompatibility of
systems. Under the modular arrangement the partners would share the same
system and use 'intranet' technologies such as Web-servers, e-mail and browser
software which is less costly and complicated than creating and maintaining a
proprietary network such as EDI.
- *Shared destiny relations* – a major barrier to shared destiny relations is the
potential for either the customer or supplier to misuse power in the relationship.
Degree of dependence is measured in relation to:
  - proportion of customer's input provided by a supplier; and
  - proportion of supplier's output to the customer.

Under the modular arrangement there is a high mutual dependency on the part of
both the customer and supplier. In each supplier category 100% of the customer's
input is provided by a single supplier and the proportion of each supplier's output
to the customer is 100%. The key influence in the exercise of power is the leverage
held by either the customer or supplier. In this case both the customer and supplier
are 'equal' partners thus minimising the opportunity for the misuse of power in the
relationship. This mutual dependency concept is clearly an important determinant
in collaborative buyer–supplier relationships. MacDuffie and Helper (1997) carried out a study on Honda's efforts to diffuse lean production concepts with
a number of its US suppliers. A striking feature in relation to Honda and supplier
dependency is the proportion of the suppliers' output to Honda. For example, in
the case of plastic mouldings and stampings the proportion of each supplier's
output to Honda was 70 and 65%, respectively.

This arrangement is quite a radical approach to supply management with clear
risks. For example, consideration would have to be given to whether the customer
would have sufficient volume to justify a single supplier factory for each supply
category. Each supplier partner would require a long-term commitment from the
customer to justify the investment involved. Moreover, in a rapidly changing
industry there is an inherent risk that another supplier outside the modular arrangement may develop a better way of doing things or develop an improved component.
Another significant impediment to an established company contemplating such an
arrangement is the likely resistance from existing employees. Unions would also be
resistant to the employees of suppliers working in the customer final assembly area
especially if their wage rates were lower. The purpose of analysing how the company

and its key suppliers would be organised under modular production is to propose a more suitable supply arrangement which reduces the risks associated with outsourcing while facilitating more collaborative buyer–supplier relationships. It has been illustrated that such an arrangement can have its merits in a number of areas such as high mutual dependency and closer collaboration. The supplier takes on greater responsibility in return for a longer-term commitment from the customer. Each supplier is part of a 'competence-based network' focusing on their distinct area of competence delivering products and services to the systems integrator. The concept of high mutual dependency is crucial to the success of a collaborative relationship between the customer and the supplier.

11.10 Concluding comments

This case study has identified the drivers and processes that can influence the outsourcing process for a company in the telecommunications industry. A number of issues associated with the outsourcing process have been considered including industry influences, business strategy, supplier capabilities and buyer–supplier relationships. When a company increasingly outsources the production and assembly of its products, the majority of activities being performed internally are the support activities of the value chain such as design, engineering, procurement, and customer service. The management and integration of these activities are predominantly people- and skill-based tasks. This case study emphasises how outsourcing presents significant organisational change implications for organisations. It represents strategic change, which in turn necessities cultural, structural and behavioural change. The evidence presented in this case study highlights the fact that to achieve the true benefits associated with outsourcing, organisations must adopt an integrated approach to the management of strategic change whereby senior managers must play a key role as facilitators of change, and those who are most affected are involved in the planning and implementation of change.

The company approached the outsourcing process from the perspective of the best available source (internal or external) being chosen to carry out each specific activity or group of activities. This process can involve a structured benchmarking approach to assessing a company's internal capabilities in a particular activity in relation to the potential suppliers' ability to provide such an activity. By accessing the most competent provider of an activity or group of activities, the company believes it is raising standards in activities that contribute to its success. The influence that the supply market can have on the outsourcing process has been emphasised. For example, it has been shown how developments in the supply market such as, suppliers developing greater capabilities, can have a significant impact upon the outsourcing process. There is considerable focus on evaluating the capability of the company in activities in relation to potential suppliers – either internal or external. This reflects the emphasis by companies in the telecommunications industry on attempting to identify and integrate a network of best-in-class product and service providers. In effect, these companies are attempting to achieve competitive advantage in the activity of specifying and integrating external services and other purchases, rather than in the assembly and production of the products themselves.

The analysis revealed a number of inconsistencies with pursuing a strategy of selecting and managing the most competent source to carry out each activity in the value chain. For example, the company used a more competent source in the logistics activity of the value chain to raise standards in this particular activity as well as contribute directly to its business strategy objectives. However, in other areas of supplier management this was not the case. Although the company personnel argued that collaborative relations with suppliers were being adopted, there was evidence to illustrate that this certainly is not the case. For example, there was clear evidence to show that the responsibility of inventory management was shifted on to suppliers in order to alleviate the problems associated with poor forecasting. However, it could be argued that the customer organisation was exploiting its position of influence in the supply chain. The type of relationship adopted with the supplier was determined by the perceived potential for competitive advantage. Rather than attempt to develop a partnership relationship, an organisation may decide that it is better to pursue a relationship where it holds the balance of power. In other words, in order to achieve the most beneficial sourcing arrangement an organisation may pursue a buyer–supplier relationship – ranging from partnership to competitive bidding – which will enable it to maximise competitive advantage. For example, an organisation may use its influence to obtain reductions in inventory and cost reductions, which in turn have a positive impact on the achievement of its own competitive position. This may also ensure greater flexibility in that the organisation will not get locked into a long-term relationship with a supplier whose technology may become obsolete. Due to the volatility associated with the telecommunications industry the duration of relationships with suppliers will depend upon how long suppliers maintain their competitiveness in terms of cost, technology and quality.

REFERENCES

Beardwell, I. and Holden, L. (1997). *Human Resource Management: A Contemporary Perspective*. London: Pitman Publishing, second edition.

Ellram, L. M. (1996). A structured method for applying purchasing cost management tools. *International Journal of Purchasing and Materials Management*, **32**, No.4, 11–19.

Ellram, L. M. and Edis, O. R. V. (1996). A case study of successful partnering implementation. *International Journal of Purchasing and Materials Management*, **32**, No.3, 20–8.

MacDuffie, J. P. and Helper, S. (1997). Creating lean suppliers: diffusing lean production through the supply chain. *California Management Review*, **39**, No.4, 118–51.

Quinn, J. B. and Hilmer, F. G. (1994). Strategic outsourcing. *Sloan Management Review*, **35**, No.4, 43–55.

Ramsay, J. (1996). The case against purchasing partnerships. *International Journal of Purchasing and Materials Management*, **32**, No.4, 13–20.

Shillingford, J. (1998). Telecoms lifts European semiconductor market. *Financial Times – Information Technology Review: London*, 4 February, p.12.

Voss, C. A. and Blackmon, K. (1996). The impact of national and parent company origin on world-class manufacturing: findings from Britain and Germany. *International Journal of Operations and Production Management*, **16**, No.11, 98–115.

12 Conclusions

12.1 Introduction

The objective of this book has been to summarise and integrate the latest research in outsourcing into a comprehensive framework for understanding the process of outsourcing evaluation and management. It has been shown that where outsourcing is evaluated and managed effectively it can be a very powerful vehicle to enhance the competitive position of an organisation. The trend towards outsourcing is set to continue as organisations are expected to do more with fewer resources. Outsourcing is not limited to the peripheral areas of the business but is increasingly impacting business areas that can contribute significantly to competitive advantage. It has been shown how organisations can benefit from accessing the supplier capabilities and employ supply relationships to enhance their own skills and capabilities. This book has also illustrated the consequences of outsourcing failure and has argued that outsourcing evaluation and management requires considerable effort and resource. Indeed, outsourcing can involve considerable risk through the loss of key skills and a failure to evaluate all the costs associated with the outsourcing process. Considerable emphasis has been placed on the issues that should be considered in evaluating whether outsourcing is appropriate for organisational activities. This involves aligning outsourcing evaluation and management with the overall strategy of the organisation. Approaching outsourcing strategically involves gaining an in-depth understanding of the competitive environment, organisational capability, supplier capability, and supply market risk. This chapter provides a number of practical lessons for outsourcing evaluation and management and also identifies a number of potential future developments in the area of outsourcing.

12.2 Practical lessons

12.2.1 The strategic context

A central theme running through this book has been the strategic nature of the outsourcing process. It has been argued throughout that outsourcing has strategic implications for the organisation and as a consequence should be at the heart of business strategy. Outsourcing can be employed to achieve performance improvements along a number of dimensions including cost, quality, service, and time-to-market. However, failure to place outsourcing within a strategic context will lead to a piecemeal approach based solely on attempts to reduce costs. Many organisations fail to establish clear strategic objectives on what they are intending to achieve and have no basis for evaluating the success of outsourcing. Rather than being approached strategically, it is often approached from a defensive position and in response to the actions of competitors. A key element of business strategy involves determining how an organisation should develop and accumulate the skills and capabilities that create and sustain competitive advantage both currently and in the future. However, a clear risk of outsourcing is that an organisation may outsource some of the skills and capabilities that are critical to future success – sometimes referred to as 'hollowing-out'. In fact, such a trend is extremely dangerous as it not only threatens the future capability of an organisation but over time can also result in the decline in competitiveness of an industry. Therefore, considerable attention and effort should be given to ensuring that the future capability of the organisation is being considered when distinguishing between activities that should be kept in-house and those that should be outsourced. Placing outsourcing in a strategic context will ensure that it is linked to the overall mission and strategic objectives of the organisation. In order to undertake this analysis, organisations have to answer the following questions.

- What are the key activities of the organisation that are a source of competitive advantage?
- Should the organisation maintain and build upon a superior performance position in a key activity?
- How much resource is required to bridge any deficiency in performance between either competitors or suppliers in an activity?
- Is it possible to outsource an activity and leverage the capabilities of suppliers?
- Will customers of the organisation recognise a difference in its end products or services if an activity is outsourced?
- In what areas of the business can outsourcing be employed to contribute to competitive advantage?

Although this type of analysis is extremely challenging for organisations, it can be very valuable. The importance and depth of this analysis has been illustrated clearly in the case of organisational capability analysis. For example, crucial strategic information is obtained in the comparison of internal performance capabilities with the capabilities of competitors and suppliers on a particular activity. In particular, the capability of an organisation in its critical activities relative to competitors is the key to building and sustaining competitive advantage. In effect, the organisation is analysing its overall capability. An organisation's competitive edge is grounded in its skills and capabilities relative to its rivals' and, more specifically, the scope and depth of its ability to perform competitively these critical activities along the value chain better than its rivals. It is also equally important to identify and understand areas of poor organisational performance. Organisations often rush into outsourcing without understanding the causes of poor internal performance. In many cases, poor performance in an area can be attributable to weak and ineffective management. Outsourcing will not improve the performance of an organisation that is under-performing due to poor management.

12.2.2   A structured approach

The process of outsourcing evaluation and management should be carried out in a structured manner. Although the analysis presented in this book has identified a number of discrete stages in outsourcing, there is considerable iteration between each of the stages. The analysis should not focus exclusively on one dimension such as activity importance, organisation capability, supply market considerations or relationship management. Indeed, in many cases the stages are linked with the analysis in one stage reinforcing the analysis carried out at an earlier stage. For example, the importance of an activity may be further emphasised if the relative capability analysis reveals that the organisation is one of a very limited number of organisations that possesses a strong performance position. A further aspect to this analysis is the conditions that are prevailing in the supply market for potential products and services. The trend towards outsourcing in both the private and public sectors has led to the development of supply markets for a range of products and service functions. Consequently, activities that may have been considered critical in the past can now be readily sourced from a number of more capable suppliers. Indeed, as has happened in many industries, companies will outsource an increasing number of business activities if a capable and competitive supply market exists. Suppliers have also become acutely aware of this fact and are aggressively developing and marketing their capabilities in order to win more business from their customers. For example, in the area of human resource management, there has been a rapid growth in the development of 'professional employment organisations' that perform most if not all human resource tasks for small- to medium-sized businesses (Konrad and Deckop, 2001).

12.2.3   The human resource implications

Outsourcing involves considerable change for the structure and employees within an organisation. Outsourcing will impact upon the employees' sense of job security, loyalty and commitment to the organisation that in turn will impact performance. Therefore, the human resource implications should be managed sensitively in order to minimise the likelihood of disruption and industrial action. The process of addressing each issue in the outsourcing process requires a participative approach between management and employees, particularly in the area of assessing the capabilities of the sourcing organisation. Outsourcing must be evaluated and managed in such a way so that staff does not regard industrial action as a justifiable reaction to the threat of outsourcing. It is more beneficial to develop a shared understanding of the activities that create competitive advantage, and an urgency around the need for improvement. Management should not use the threat of outsourcing to intimidate employees if performance is lacking in certain areas. For example, the threat of outsourcing is often used as a weapon to secure salary reductions from employees. Outsourcing is often considered in response to the need to transform the performance of a poorly performing area of the business. In certain circumstances, it may be more prudent for management to give employees the opportunity to improve in order to meet external benchmarks.

12.2.4 Supply management capability

Outsourcing does not eliminate the need to manage an activity once it is outsourced. Rather, it creates a situation requiring managers to possess and utilise a different set of skills. Outsourcing often involves significant changes in the way in which an activity is managed. Control is exercised through a contract rather than through direct ownership of assets and employment of staff. This involves managers of the outsourced activities selecting, negotiating with and managing vendors. Therefore, prior to outsourcing an activity, organisations must ensure that these managers have the skills to assess whether suppliers possess the required capabilities and a competitive supply market exists. In many cases, managers lack these skills and more importantly are unaware that managing an external supplier requires a different set of skills from those associated with managing an internal function. Organisations should recruit staff who possess the necessary skills or invest in training and development for its existing staff who will be charged with managing the relationship with suppliers. For example, this may involve the sourcing organisation recruiting managers with previous experience in successful outsourcing strategies. As well as possessing supplier management skills, the sourcing organisation must retain a level of technical know-how related to the outsourced activity in order to assess current and future technological developments. The managers responsible for managing the outsourced activity should develop and implement a strategy in order to ensure it is aligned with the overall business strategy of the sourcing organisation. This will involve considering both current and future supplier capabilities. For example, in many supply markets the complexity of the demands by organisations has posed challenges and in some cases outpaced the capabilities of suppliers.

The ability of organisations to manage suppliers is a key element of any out-sourcing strategy. As the trend towards outsourcing more important business activities continues, the approach of companies to supplier management will become even more critical. The organisations that will gain most from outsourcing will be the ones that can unlock the potential opportunities that exist in supply markets. The case of the Japanese car makers in the US has illustrated that it is possible for some companies to obtain much higher performance levels from suppliers than their competitors. Essentially, these companies are utilising a resource (i.e. the supplier) more effectively than their competitors. In order to achieve this objective, organisations must have a basis for allocating resource to managing particular outsourcing supply relationships. A portfolio approach to outsourcing supply relationships is required, with each appropriate relationship type delivering a distinct set of benefits to the sourcing organisation. For example, through employing a short-term adversarial relationship, the sourcing organisa-tion has the opportunity to access many suppliers and rapidly switch suppliers with little cost or inconvenience if it is found that another supplier is more competitive. Alternatively, the sourcing organisation can employ a long-term collaborative relationship as a means of supplementing the skills and capabilities of the sourcing organisation both currently and in the future.

12.2.5 Performance management

Performance improvement is typically the most often-cited motive for outsour-cing. Therefore, one would assume that organisations would give considerable attention to both identifying and measuring the short- and long-term performance gains derived from outsourcing. However, organisations often experience diffi-culties with performance measurement. Performance objectives and measurement criteria should be established at the outset of the outsourcing relationship. For example, the motivations for outsourcing are often the pursuit of service-related improvements. Clearly, the sourcing organisation must establish performance measures and a formal system to ensure that such performance objectives are being met. Employees in the sourcing organisation must also be clear on what the objectives of outsourcing are and how they are aligned with business strategy. This performance system should also be an integral part of the contract management process. Effective contract management requires reliable information that can assist in tracking performance. Although the complexity of the performance management system will be related to the nature of the contract, the effectiveness of the outsourced activity must be assessed. In some instances, contracts involving expensive and complex services may require evaluation mechanisms designed specifically for that one contract. Performance measurement should not be focused solely on the outsourced activity. For example, many organisations outsource peripheral activities from certain functions in order to allow the function to focus on more strategic activities. It is important to ensure that this objective is achieved through analysis of the performance of the internal function subsequent to outsourcing. Therefore, attention should also be given to the impacts of outsourcing on the internal function rather the external supplier alone.

12.3   Future developments

The practice of outsourcing is already well developed in the private sectors of many countries. However, recent studies have shown that different regions in the world are at different stages of development in the practice of outsourcing. Kakabadse and Kakabadse (2002) have found that US companies have been pursuing more value-adding outsourcing strategies through the pursuit of improved service quality, a greater focus on business activities that are crucial to success, better use of technology, whilst at the same time reducing costs. Alternatively, European companies have been focusing more on achieving scale economies and cost savings through outsourcing. Although outsourcing is gaining prominence in the public sector of many countries, it still has some way to go before it reaches the level of development that is seen in the private sector. The increasing use of outsourcing in the public sector has been directly attributable to the introduction of the free market ideologies of the Thatcher and Reagan governments in the UK and US during the 1980s. One might assume that these ideologies and associated policies would have become less prevalent with changes of governments in these countries. However, this has not been the case. For example, in the UK many of the policies associated with the free market ideology have endured throughout the Blair years. Many governments have continued to view outsourcing as an appropriate way of achieving performance improvements in services and greater choice for their citizens. However, the extent to which public sector organisations can pursue outsourcing relative to the private sector is and will continue to be constrained by the presence of political influences.

Chapter 2 identified a number of outsourcing drivers including globalisation, developments in information and communication technologies, public sector reforms and more demanding consumers. Although the pace of outsourcing is difficult to predict, these drivers will continue to influence the trend towards outsourcing by organisations in both the private and public sectors. These trends in many cases are inter-related. Indeed, the likely development and progression of outsourcing will be influenced by the success of current outsourcing efforts. Some of the potential future developments are given here.

- *More demanding sourcing organisations* – organisations are already dealing with an extremely sophisticated and demanding consumer. The increasing level of competition and the further development of the Internet as a channel for communication and trading in many markets will continue to further empower the consumer. The growing product and service complexity demanded by consumers is likely to further the trend towards outsourcing. The increasing demands of consumers on organisations are feeding through to the suppliers of these organisations. Although many organisations are increasingly outsourcing more critical business activities, most organisations currently outsource only peripheral business activities. As organisations outsource a greater range of activities and a higher level of value associated with each activity, suppliers will be expected to provide a broader range of services. Rather than employing outsourcing to solely deliver cost reductions, more organisations will outsource activities in order to obtain a higher level of value from suppliers in areas such as greater responsiveness and flexibility. The focus will move from a single functional perspective towards employing outsourcing as a means of reconfiguring a group of organisational activities. However, a danger of outsourcing a broader range of activities is that suppliers will be unable to keep pace with the demands of customers. Organisations will be forced to place more emphasis on evaluating supplier performance and supply market developments.

- *Greater outsourcing complexity* – outsourcing evaluation should be relatively straightforward in the case of peripheral activities. However, as organisations increasingly outsource more critical business activities, the processes of selecting suppliers, negotiating contracts and managing the transfer of assets and operations to the supplier are likely to become more complex and costly to implement. Contracts will become more sophisticated in terms of performance measurement mechanisms, asset transfer management and the development of clauses to allow for outsourcing failure. The tendency to use more external suppliers will become more tempting as many supply markets develop to offer a wider range of business services – both locally and offshore. Indeed, many organisations will reach the point at which almost all the activities required to produce and deliver their product and service portfolio can be performed entirely be external product and service providers. A key element of business strategy is determining the scope of organisational activity that can create competitive advantage both currently and in the future. In fact, the process of outsourcing evaluation and management has become a critical activity for the organisation. Developments in information and communications technologies are further fuelling the move towards the adoption of alternative relationship configurations such as network organisations and virtual corporations. As well as accessing the capabilities of suppliers, organisations will continue to employ outsourcing as a means of reducing financial risk in areas such as new product development. For example, in the past, many of the large carmakers shouldered most of the burden of component module research and design expenditure for new models. However, they are now encouraging suppliers to invest resource in research and design in return for longer-term contracts and more business.

- *Increased offshoring* – although some organisations have experienced difficulties with offshoring certain activities in the areas of service and quality levels, this trend is set to continue. This trend is likely to be more pronounced amongst private sector organisations than public sector organisations. Indeed, legislation has been enacted in some countries such as the US, in order to prevent the outsourcing of public sector activities offshore. However, many private sector organisations will continue to seek offshoring opportunities as they can achieve substantial cost savings and in some cases productivity gains. Many offshore locations for product- and service-related activities are aggressively marketing their regions to both private and public sector organisations in the developed economies. India is no longer the main source for outsourcing services with other countries competing for these services with for example, China, Malaysia and some parts of Eastern Europe increasingly becoming suitable locations for such services. The governments in these countries are employing strategies to enhance their attractiveness to outsourcing organisations in more developed economies. Furthermore, some of these countries are experiencing considerable growth in their outsourcing services sectors with many of the companies offering a broader range of services in order to encourage their customers to outsource a broader range of activities.

- *The changing organisation–employee relationship* – organisations are increasingly supplementing their traditional structures with a range of different working arrangements including external consultants and temporary workers. A study undertaken in the UK found that around 90% of organisations outsource one or more services and around 60% of organisations employ both subcontractors and one other form of non-standard labour (Cully et al., 1999). As a result of the use of more flexible working arrangements, the relationship between many organisations and their employees has radically changed. Increasingly, many individuals have been specialising in an area of expertise and are no longer loyal to a single organisation. In turn, the paternal role of the organisation in its relationships with its employees has been gradually disappearing. The development of specialist organisations in outsourcing supply markets has presented opportunities for individuals to develop specialist expertise and experience. Research on information technology professionals in the US has found that contract arrangements in some cases increased skill levels through experience of different workplaces, friendships and collaborations (Grugulis et al., 2003). A major effect of this trend has been the shift of responsibility for career development from the organisation to the individual. Due to the lack of job security and the trend towards employee externalisation, individuals will have to develop knowledge and skills that allow them to contract with successive employers. A concern with this continued trend is that these conditions are not conducive to allowing individuals to develop their knowledge and skills base. For example, Mallon and Duberly (2000) found in a study of career professionals employed on a consultancy basis that these individuals were still relying on knowledge gained at the start of their careers. Under the paternal type organisation skills development was supported through a range of mechanisms such as training, promotion, mentoring and development. Although organisations are seeking greater flexibility and lower costs through outsourcing, a concern is that they are reducing the skills base available in their respective labour markets. Furthermore, this trend has wider economic implications as it has the potential to create a spiral of low wage and low skills within an economy.

Much of the analysis in this book has been at the firm level. However, some of the analysis presented is also relevant to the wider economic environment and in particular international competitiveness. Organisations in their approach to outsourcing have essentially been attempting to focus on critical activities that create competitive advantage whilst outsourcing peripheral activities to more competent external suppliers. In effect, a similar trend has been occurring in many developed economies, as many manufacturing- and service-related activities have been increasingly outsourced to offshore locations. Granted, much of the impetus behind this trend has come from organisations as they strive to produce and deliver lower cost products and services in response to increasing competition. However, although governments in developed economies have not actively encouraged companies to transfer much of their manufacturing or service activities offshore, many governments have failed to act for example, through protectionist measures, in order to prevent such a trend. In fact, as evidenced by many of the policies being adopted in developed economies, governments recognise that their home-based companies cannot compete in many areas of business activity with the low cost bases of some offshore locations. For example, industrial development agencies in these countries are no longer willing to fund areas that are at risk of being offshored but focusing on offering incentives to more knowledge-based areas of their economies. Indeed, a major challenge for developed economies is to increase the level of added value and innovation capacity in order to compensate for the decline in jobs as a result of offshoring in manufacturing and service areas. However, a major concern for developed economies is that high value-added jobs are also under potential threat from offshore locations. This is a justifiable concern as many developing economies continue to develop their skills base through investment in training and higher education. In the same way organisations have to determine the scope of business activities they perform internally, governments are also being challenged to determine the areas on which their economy should focus in order to enhance their long-term competitiveness.

REFERENCES

Cully, M., Woodland, S, O'Reilly, A. and Dix, G. (1999). *Britain at Work: As Depicted by the 1998 Workplace Employee Relations Survey*. London: Routledge.

Grugulis, I., Vincent, S. and Hebson, G. (2003). The Rise of the 'network organisation' and the decline of discretion. *Human Resource Management Journal*, **13**, No. 2, 45–59.

Kakabadse, A. and Kakabadse, N. (2002). Trends in outsourcing: contrasting USA and Europe. *European Management Journal*, **22**, No. 2, 189–98.

Konrad, A. M. and Deckop, J. (2001). Human resource management trends in the USA: challenges in the midst of prosperity. *International Journal of Manpower*, **22**, No. 3, 269–78.

Mallon, M. and Duberley, J. (2000). Managers and professionals in the contingent workforce. *Human Resource Management Journal*, **10**, No. 1, 33–47.

Index